TOMORROWLAND

TOMORROWLAND

THE FUTURE OF THEME PARKS

MADISON KELLEY

NEW DEGREE PRESS

TOMORROWLAND

The Future of Theme Parks

ISBN   978-1-64137-034-9   *Paperback*

ISBN   978-1-64137-035-6   *Ebook*

CONTENTS

INTRODUCTION

———

"Remember, you can use your muggle cameras!"

As we waited to enter the door before us, I heard laughter and a loud series of screams from a crowd gathered down the street from us. I felt the heat. "Must be the dragon breathing fire at Gringotts," I thought to myself.

I looked up above my head, where a hanging sign swayed subtly outside the building. It said Ollivanders in a script-like font, with what looked like a wand making the O and then jutting out over the rest of the words. Underneath was written Maker of Fine Wands since 382 BC.

"Again, you can use your muggle cameras," the man said to the group, this time louder to try being heard over the gasps and screams from down the street. Then he looked at me and said, "Muggle cameras that take pictures that don't even move," smiling and rolling his eyes slightly.

"Be respectful to the wand master," he continued, looking directly at the group of four first-graders dressed in black robes, each bearing matching patches with a lion that read Gryffindor. "Please, no flash," he said, this time directing his gaze to the two middle-aged women holding their iPhones poised for the next photo opportunity.

I was nervous; my palms were sweaty. It was a feeling unlike anything I'd felt in my normal muggle life. My heart was starting to race a bit, and I smiled to myself and thought, "This is okay, right? I'm not too old for this, right?"

I was standing at the front of the group, which meant I'd be one of the first to be escorted to the wand master's office—I would've been *the* first, but I figured I should let a couple of these six-year-olds with us go first to be polite. As I walked down the long corridor, seeing floor to ceiling, row after row of wands in boxes of different colors and textures, I forgot about my life and any outside troubles. My world was slowly melting away behind me. I never looked back. I wanted to go forward. I wanted more. I never wanted to leave this world and return to my own reality.

We arrived at an open door that led us into the dimly lit, slightly musky room. The assistant who had been leading us down the corridor extended his arm to the corner of the room, gesturing for us to stand there. The room was surprisingly large. At the front of the room, there was a big desk with a shelf full of more wands behind it. There was a man standing behind the desk, scribbling something on paper. His long beard swayed slightly as he wrote each letter. To my left, there was a staircase leading up to bookshelves with more wands and other books. There were other trinkets scattered around the room, like a flower vase, books, and a quill, but I could barely focus, trying to take everything in. As the rest of the group shuffled in, I felt the room filling with humans and hushed voices. The assistant motioned the young and the short to the front, while tall parents and others were motioned to move toward the back.

"Thank God I'm short," I thought to myself.

As the man who I assumed was the wand master laid his quill pen down on the desk, picked up his spectacles, and put them on his nose, I took a deep breath and lightly bit my lower lip.

I opened my eyes wider in an attempt to capture every detail of what was about to happen.

The wand master began to move from behind the desk, folding his arms in the sleeves of his long robe.

The images of Harry's wand selection in the first movie flashed in my head. And a refrain played in my head: "Pick me! Pick me! Oh please!! Pick me!"

I was concerned that someone's phone would ring during the ceremony and ruin it for all of us. So, I scanned the room quickly to see whose phones were out.

Then, I locked eyes with the Wand Master. My heart stopped.

* *

We *love* our phones. Just how much, you'll never guess. A recent study found that

- 43 percent of iPhone owners were willing to go barefoot for a week rather than give up their phone for a week;
- 33 percent of smartphone users were willing to give up sex for a week rather than forego their mobile phone;
- 55 percent were willing to go without caffeine, 63 percent said they'd give up chocolate before their phone, and 70 percent would forsake alcohol to hang on to their phone;
- And here's the one that *really* got me—22 percent of smartphone owners said they would rather go a week without seeing their **significant other** than without checking their apps.[1]

So yeah, we really do *love* our phones.

Nowadays people find that not only can they not live without their phones, but also, they are increasingly living within them. Conversations and social activities are happening over FaceTime and Facebook instead of face-to-face. People's instincts are to email or text someone before calling them. Think about the way business used to be conducted. Everything happened over the phone or in person. Now, if people want to discuss a small business matter, they send a quick email instead of picking up the phone, because it's quicker and easier.

Even in places that are inherently "in person" suffer the same consequences. People are watching concerts, solar eclipses, and even the Disney parade through their iPhones. The screens have become their eyes, watching what is happening right in front of them through their new lens.

It's unlikely to change anytime soon; in fact, it will probably get worse. Because social media and the internet allow us to access an abundance of information quickly, our attention spans have dropped. The average attention span in 2000 has now gone from twelve to eight seconds and is only continuing to dwindle.[2]

And because information comes to us instantaneously, since it takes two seconds to Google any question, we expect the rest of the things in life to come just as quickly. (Isn't it intolerable when your iTunes movie has to download for 20 minutes instead of just playing immediately?) While I might disagree that our addiction to social media will only further entrench us in media technology, I do agree that the effects of technology on our daily interactions with the world will not change.

However, for that moment in the presence of a Wand Master in Ollivanders, technology was the furthest thing from my mind. I'm a millennial, and yet I never felt the urge to pull out my phone. My visit that day to Universal Studios' Wizarding World of Harry Potter was pure. I felt childlike wonder. I was present.

But it was also rare.

I ask myself, "Will technology—and our smartphones—eventually alter the pure, unadulterated experience I had in Ollivanders wand shop? What happens when we all have augmented-reality glasses and can be checking our text messages without having to pull out our phones? Will our brains be conditioned to need more in order to feel experiences? Has the overload of more, more, more ruined our need for Harry Potter World? How long until we don't even bother to come to the park and experience it all through VR goggles or can just become a character in the movies ourselves through our gaming systems?"

Will technology change the theme park? Could it actually make them irrelevant?

*   *

"Theme parks are dying," my brother Jordan said. He is a startup founder in San Francisco, and he—like many of his friends—was quick to point out how technology will disrupt everything.

I remember years ago, before I'd ever ridden in an Uber, when he said that taxis would die, and when he said that the hotel industry would change forever with Airbnb, and when he said that newspapers were a dying industry.

And to some extent, he's been right. Uber is a verb. Airbnb is effectively the largest hotel chain on the planet. And Facebook receives more in ad revenues than the entire newspaper and magazine industry, *by itself.*

Then is he right about theme parks? Are they next?

His comment shocked me—especially as a person with my level of enthusiasm toward this particular industry.

"No," I said back quickly, furrowing my brow as I'd done with my older brother since we were little kids. "Why would you ever think that?"

"Technology is becoming so realistic and immersive that you don't need to go to humid, muggy Florida to wait in lines and pretend to talk to humans in costumes," he said. "Eventually you'll put on a headset and haptic gloves that make you feel like you are walking or moving. Then you'll plug in, and you don't have to put up with all the strollers and crap of a theme park. Heck, you can fly from ride to ride. You can do whatever you want. Why would anyone bother to go?"

As much as he sometimes just says things to rile me up, I began to wonder if he was right. Would technology eventually kill the theme park?

Today his prediction looks far-fetched, as the theme park business is booming. In 2016, 148 million Americans visited the top 20 theme parks. In 2013, it was 135 million Americans. Even between 2015 and 2016, overall attendance went up 1.5 percent.[3] But if technology gets good enough to replace the feeling of going to a theme park, could those numbers begin to reverse?

* *

People believed that technology wouldn't be able to help us because it would end up as virtual visits only, but applied in the right ways, technology could make theme parks more amazing.

—MOHAMED NEWERA

Google the phrase *VR roller coaster*, and you will find apps on Google Play and iTunes, and other apps for VR headsets, that allow you to experience riding a roller coaster right from the comfort of your couch. While the reviews are mixed, some of the applications appear to be quite realistic even in the infancy of technologies like virtual reality.

Why bother spending hundreds of dollars at Disneyland when you can ride the Matterhorn, enjoy a pickle, and paint your own face at home?

Having spoken to dozens of experts and executives in the theme park and entertainment industry, I have come to a very different conclusion about the future of theme parks in an era of technology—particularly increasingly immersive and realistic technologies.

Technology will make theme parks *better*.

Theme parks have always been on the bleeding edge of technology. Think about Walt Disney's fascination with Epcot and Tomorrowland. Today rides are already embracing technologies to make the rider feel more immersed, and new exhibits from Universal and Disney have led to a renaissance of parks that, to some, had gotten stale. We are entering the golden age of the theme park, and the technology my brother thinks will disrupt the industry altogether is actually going to make theme parks more popular than at any point in history.

Why do I say this?

As I've written this book, I've been floored by the responses from industry experts I've spoken with:

- I talked to Imagineers who have been in the industry for 20 years, everyone from the engineer to the conceptual designer.
- I interviewed the virtual-reality, augmented-reality, robotics, and numerous other emerging-technology experts to understand how technology will eventually integrate into the theme parks themselves.
- I researched, read, and spoke with many industry experts to learn about the history and the future of the business of theme parks.

And much like me, the people in and around the industry see these new technologies enabling the next great theme-park evolution rather than removing the need for their existence.

In fact, more technology actually has made us crave technology-free (or technology-light) activities. Studies have shown that young generations, born between 1995 and 2012, are happier when they are off their phones and doing off-screen activities.[4] This is not to say that the future of theme-park attendance will be strictly tech-free, without Snapchat filters or Instagram locations. Instead, people just want time to be with each other, in person, doing activities in real time, and one of those activities will continue to be attending theme parks.

*　*

This book is meant to help all of us learn about the future "Happiest Places on Earth." This book offers some ideas that may not even be in the imagination of future theme-park designers, based on emerging technology still in its infancy. Inside this book, here's what you'll learn:

1. How to be Alice and Han Solo in the same weekend.
2. How you won't have to wait in lines anymore.
3. How you can meet your new tour guide: Siri's great-great-great granddaughter.
4. How you won't be interacting with felt-suit characters anymore but the next best thing.

Here's the big idea: ***Theme parks are here to stay, and they will only get better.***

And while I am writing this book mainly for the theme-park

geeks like myself, I also wrote it to be a "how to play with each other" book for the technology developers and theme-park developers who together can create the next great attractions.

Many of the stories you'll read are firsthand experiences of the rides and the parks themselves from the perspective of a customer. They are experiences somewhat similar to your own, but now seen through the lenses of theme-park designers and engineers, and today's leading technologies, you'll get a picture of the unique perspective I know when I step across the yellow line. You'll hear stories about

- How I realized I was super out of shape when I went to Alton Towers;
- My reactions to riding Escape from Gringotts at Universal Studios. Some are funny, and some will make you cry, as I did when I experienced them and wrote them down;
- What it might look like to stay in the Star Wars Hotel; and
- Why I love theme parks so much and why, as a grown woman, I am still going. I will be that 80-year-old with her face painted, screaming in bliss on the Hollywood Rip Ride Rockit at Universal Studios.

So while this book is written for theme-park goers and fans, I hope it offers industry professionals a unique and different view of technologies and the potential for impact on theme parks. Even I was blown away by the technology that will rapidly become mainstream. As we think about these innovations in Silicon Valley and their power to transform whole industries, I am excited to imagine how it might enhance theme parks. Whether you're the CEO, a designer, or the general contractor, this book will offer a unique look at the future. While I won't delve too deeply into the actual planning of the park or how to construct the experiences I discuss in the book, instead it will be sort of a crystal ball into the technological possibilities of the present and the foreseeable future. This book, though informative for the theme-park employee, is also a book that I would hope any sci-fi nerd would get jazzed about. (I mean, you can't have a book about theme parks and have it be a snooze to read.) Although this book may be for the veterans of the industry, I have eliminated the technical jargon that both the tech side and the theme-park side use, so the book is more fun to read. I am a fan. This book is for people like me, and if you picked up this book, then it's definitely for someone like you.

These are the big trends that we are seeing for the future of theme parks:

1. Increased use of immersive technologies like virtual and augmented reality.
2. The more immersive the better. Lands like Harry Potter, Pandora, and Star Wars will become the new norm.
3. "Nothing kicks ass harder than a good story."

You should read this book because it is going to be like stepping into the world of *Her* and *Westworld*. If you decide to put this book back down on the shelf, you'll miss out on the experience of walking through the newest theme park: Technoland, where you can live the day as anyone you want to be.

Oh, and in case you're wondering about what the wand master at Harry Potter World selected for my wand...

Well you'll just have to read on to find out.

PART I

CHAPTER 1

CHASING THE HAPPY

———

"Welcome. To. Ollivanders," the man bellowed in a grizzled, British accent. "I am the wand master."

My heart was beating. My mind was chanting: "Choose Me! Choose Me! Choose Me!"

Dressed in a long gown, his arms folded in his sleeves, with a cloak thrown over his shoulders, the wand master walked up to the woman three people down from me and said, "You!" pointing at her, with his long fingers wearing a bevy of jeweled rings.

He continued, "You've come to find a wand, have you?"

She nodded, and I watched as she visibly stood straighter and taller.

At the same time, I was heartbroken. I felt a sense of annoyance rush over me. I wanted to be the person who was speaking with the wand master. This was not how it was supposed to go.

The wand master took the chosen woman by the hand and led her from the group. They walked to his desk, and he stood her in front while he walked behind it. He paused and stared at her—deeply looking at her face, her hands and back to her face.

There was a silence in the room, and I—like the rest of my group—was transfixed by the wand master. My previous feelings of frustration began to fade, and I was part of the scene. In many ways, I felt like I was her.

The moment passed, and he quickly turned around with his robes flowing in an arc around his body as some of his tassels bounced off the desk. He paused as his back was turned away from us before he reached for the wall behind him.

"Ash," he bellowed. He grabbed a wand from the wall, turning back to our group and the woman standing in front of his desk.

"This is a wand of ash. It clings to its true owner and will not have the desired effects for the wizard to whom it does not belong. Do you understand?" he asked gravely.

She nodded before he placed the wand into her outstretched hand.

"My flowers look like they could use a watering," he continued, motioning with his arm toward a pot of flowers in full bloom.

"Now point your wand at my flowers," he said, motioning toward the flowers, "and say 'Aguamenti!'"

The woman paused, looking like she was saying the word to herself first before blurting out the spell, and finally shouted, "Aguamenti," with a heavy emphasis on the last *Ee* sound.

The room darkened slightly, and a crack of thunder shocked all of us as we leaned back away from the scene. The wand holder was visibly frightened and dropped her arm holding the wand to her side.

Before our eyes, the flowers that were in bloom in the pot on the shelf above the staircase to the left began to wilt and die, until they were limply lying over the edges of the pot.

"Oh dear!" he said as he let out a breath, expressing some concern.

"This," he continued, "is definitely *not* your wand. But no worries."

With a flick of his wrist toward the wilted flowers, we saw his tulips return to perfect color and health. I felt my heart quicken. "That was cool," I thought to myself, as any earlier feelings of annoyance had completely disappeared. Maybe some of that was due to the fact that she hadn't found her wand on the first try. "A little jealousy is okay," I thought to myself.

"Hmm," he said with a wave of his finger as his other hand stroked his long, white beard. "Aha! I've got it."

He turned around to the wall behind him, with his robes again making an arcing circle, and reached for something I couldn't quite see from my position. He turned back toward the woman and the group, this time holding a different wand from the shelves behind him.

"This!" he said triumphantly, raising the wand for the entire group to see. "Is a wand of hawthorn."

The wand was longer and made of a lighter-colored wood. It reminded me a bit of the wand Hermione Granger had—I know that wand when I see it.

The wand master continued, "This particular wand is suitable for healing."

Suddenly the room flashed as lightning could be seen through the window in the small room.

"Worry not!" said the wand master. "I can fix this!"

With another flick of his wrist, the lightning stopped. He didn't say anything this time as he reached behind him for the last time.

*　*

"Disneyland is your land. Here age relives fond memories of the past, and here youth may savor the challenge and promise of the future. Disneyland is dedicated to the ideals, the dreams, and the hard facts that have created America…with hope that it will be a source of joy and inspiration to all the world."—Walt Disney

The American Dream can be boiled down to the idea that people want happiness and security. I was interviewing Scott Acton, the CEO of Forte Specialty Contractors, a special contracting company hired to design experiences in everything ranging from restaurants to nightclubs, and they even redesigned a theme-park attraction. Scott told me an interesting fact that I previously wasn't aware of: When the economy is down, theme parks are still up. I found that very strange because, to me, it would make sense that when the economy is down, people want to save their money instead of spending it on a day at Disneyland. But the more I thought about it the more it made sense.

People, in times of turmoil, want to escape. They want to find some time to smile and be carefree, even if just for a moment or a few hours.

Disneyland was founded in the aftermath of the Great Depression. The American public not only had more money to spend, but they also looked to escape the horrors they just lived and look forward to a more hopeful future. Walt Disney himself was looking to escape his failures as a businessman, so he created this perfect little world where he could control and manipulate until it was "perfection," as Salvador Dali described it. "When you think of a man like Walt Disney as the guy who's always looking to the next horizon...and then he can say, publicly, 'I'm never going to make anything as good as *Snow White*,' that's a man in crisis."[1] At this point in Walt's life, he was feeling out of control, and Disneyland was, "comfort and salvation, and a working surface for the disappointments and confusion that [came] to him." The creation of Disneyland was a turning point for Walt. He became so invested, in a way that he hadn't felt since *Snow White*, so much so that during the planning phases, he was in Anaheim almost every day. And after the opening of Disneyland, it's said that he would get up in the morning, before all the guests came in, and in his bathrobe would go to the store that sold fresh-squeezed orange juice on Main Street and bring some back to his apartment above the fire station for breakfast.[2]

The reason that Disneyland still strikes a chord with Americans in particular is because it points out all the good in our past and looks toward a hopeful future. Americans flocked to Disneyland on a "simple promise: a day's escape from the cares and concerns of everyday life."[3] So Disneyland's separate worlds represent that escape, Fronteirland as the past, Tomorrowland as the future, and Fantasyland and the magical alternate plane representing a better, more distant present.

Disneyland would be a world of Americans, past and present, seen through the eyes of my imagination—a place of warmth and nostalgia, of illusion and color and delight." "Physically, Disneyland would be a small world in itself. It would encompass the essence of the things that were good and true in American life. It would reflect the faith and challenge of the future, the entertainment, the interest in intelligently presented facts, the stimulation of imagination, the standards of health and achievement, and above all, a sense of strength, contentment, and well-being.

—WALT DISNEY

The American frontier was founded on the idea of freedom from the past and from restraint. Frederick Jackson Turner's thesis on the American frontier focuses on how the frontier developed the idea of democracy so central to American culture. He argues, "That coarseness and strength combined with acuteness and acquisitiveness...that dominant individualism bred by expansion" is the core of American culture. He furthers, "American democracy was born of no theorist's dream; it was not carried in the *Susan Constant* to Virginia, nor in the *Mayflower* to Plymouth. It came out of the American forest, and it gained new strength each time it touched a new frontier."[4]

Disneyland's Frontierland is a direct homage to Turner's thesis. Walt Disney was looking back to a past, but "not the complicated moments that are about pain and suffering," instead the moments of resolution and exploration that shaped our American present, and now even our future.[5] When Disneyland first opened, diplomats from all over the world, from the Shah of Iran to the King and Queen of Nepal, made a stop there because to them, it was, "the CliffsNotes version of American history and culture." "Soviet Premier Nikita Khrushchev threw a fit when the US state department cited security concerns and quashed his planned visit to Disneyland."[6] Fronteirland and Disneyland in general are a microcosm of American culture and history.

* *

"This wand," he said in a hushed tone as he opened the box to reveal the wand, "is a wand of reed, best suited for the bold wizard and an eloquent speaker."

He paused, turning the wand over in his hand for all of us to see. "This wand has a core of... dragon heartstrings."

A barely audible gasp came from our group. We knew this wand was special.

"Your loyalty will be greatly admired among your friends."

He handed the wand over to her, and before it even reached her hands, the lights started to flicker. The second the wand was in her hand, a spotlight hit her, wind picked up, and we could hear the *dum-dum-da-dum-da-da-dum* of the *Harry Potter* theme song.

Goose bumps formed on my arms.

This.

Was.

Incredible.

I shook my head in disbelief as to what had happened in front of me.

I was 85 percent amazed and 15 percent jealous of this girl who just had the most amazing experience of her life, or at least of mine.

And here's the craziest part: *I wasn't even the wizard who was chosen.*

* *

Admittedly, I am an *avid* theme-park goer. If I periodically appear like a crazed fan, it's probably because I am. But this book is really intended to share the wonders and experiences of both an avid fan and a casual admirer of theme parks.

In October 2017, I visited Universal Studios for the second time. I have visited Disneyland at least once a year for the past four years. And every time I go, my experience is different. Something has changed to make it better: a new ride or exhibit; a special interaction with a character; a tip on how to avoid the lines, what pickle stands to go to, or which rides to ride when. Because I care, here are some of those tips:

- I've learned that when the park opens early to the hotel guests, I should immediately hit Space Mountain, because it becomes way too crowded later in the day. It would have been nice back then to have the Disney app that shows the wait times.

- I've learned that while you're waiting in line for Indiana Jones, it's best to grab a pickle from the cart next door, because Disneyland has the best pickles.
- I've seen the power of the Disney MagicBands, meaning I don't have to carry cash and worry about it falling out of my pocket.
- I've learned to *always* have the Express pass at Universal Studios. Express passes are basically fast passes, meaning you can skip the regular line for an expedited one, but unlike the Fast passes, you can use them at any time and on every ride.

Every time I go to Disneyland, Universal, or any theme park, I think about my own experiences and how the millions of others experiencing this with me could improve, enhance, or be different. And as the cool technology like the apps and the MagicBands came out, I began to see elements of technology that take these experiences to the next level.

*   *

All of us have our own inner child—the little girl or boy who, like Peter Pan, never wants to grow up. If I had a terrible day as a child, whether the kids at school were mean or I got into a fight with my parents, I would always turn to books or movies. Being able to dive into someone else's life, someone who had way bigger problems than I did, helped to put my life into perspective and, most of the time, made me forget all the bad things that had happened that day. When I was reading *Harry Potter*, I would imagine myself as part of Harry's friend group.

Even becoming an adult, I've never quite lost that wide-eyed curiosity. I waited for my Hogwarts letter to arrive until I was accepted to college. The magic these stories brought to my boring life made me believe that anything was possible. When I first visited The Wizarding World of Harry Potter, I looked up, saw Hogwarts castle, and said to myself, "I've finally made it." I didn't need my acceptance letter anymore, because for *that* day, I was Hermione Granger—and I had the wand to prove it.

*   *

We're all chasing the happy. The key to life is happiness, right? And we'll spend our whole lives searching for it, in ourselves, in our work, in each other. But Walt found happiness in looking back, to childhood, to the innocence that we all felt at one point, and he gives that back to us.

When I went to Disneyland for my twentieth birthday, I really felt like a princess. Every single park employee who saw my birthday button told me, "Happy birthday!" I couldn't walk down Main Street without hearing "Happy birthday!" My friends were so over it, but it made me smile every time because it made me feel like a kid, when the whole world stopped on your birthday.

Theme parks let each of us access parts of our childhood that get lost in the mundane day-to-day of studying and work life. I love the movie *Alice in Wonderland* because it's the story of a girl who feels out of place in her life, and she learns, in Wonderland, to revel in the person she is. And when I get my face painted at Disneyland, and I get some weird looks because I'm definitely supposed to be "too old" for that stuff, it gives me the confidence I had when I was young, to assume that it's just everyone else around me who's weird and not me. So I strut my stuff, with a sparkly unicorn painted on my face, and I eat nothing but ice cream and churros, and I smile nonstop.

The nerd in me just wanted to put pen to paper about all the fantastic things technology can do to enhance the theme park experience, not to eliminate it entirely. I believe that technology, like virtual and augmented reality, can make the experiences that I had at theme parks more realistic and more immersive. I just want to bring the joy that Disney and Universal brought me to other people.

CHAPTER 2

BLAST FROM THE PAST!

———

TIME TRAVEL THROUGH
THE HISTORY OF THEME PARKS

———

It's a world of laughter
A world of tears
It's a world of hopes
And a world of fears
There's so much that we share
That it's time we're aware
It's a small world after all
It's a small world after all
It's a small world after all

It's a small world after all
It's a small, small world.

Think back to Walt Disney—a man who invented some of
the most iconic cartoon characters in history. While today
we think of Walt as one of the most genius innovators in
history for his ability to bring characters to life and ultimately
translate them into the most iconic theme park in modern
history, many of us forget that it was a pretty incredible risk
for a cartoon illustrator to create a theme park.

Can you imagine what his first investors must have thought?

"Let's trust a cartoon illustrator to build a theme park."

"We have just 12 months to construct the vision of this single
man on 160 acres in Southern California."

"He's done wonders on television, but why would he want to
create a fancy carnival?"

And in truth, these concerns were why Disney struggled to
get traditional financiers for his venture. What those potential investors might have failed to realize was that Disney's
creative mind coupled with his experiences visiting amusement parks with his daughters in the 1930s and 1940s gave

Walt the palette he needed to design one of the most remarkable creations in history:

Disneyland.

While this and Walt's subsequent creations serve as iconic representations of theme parks, Disney didn't *invent* the theme park. To get to the start of theme parks, we need to go as far back as the 1500s and see some of the early steps that would serve as the foundation for Disney's creation.

Let's start by making the distinction between amusement parks and theme parks. Amusement parks tend not to have a themed aspect and are more or less just a compilation of entertainment and rides. If the park has a myriad of roller coasters without a single tying theme, and you think, "Wow, this doesn't go together," that's an amusement park.

On the other hand, the theme park is built around a larger idea that is present throughout the entirety of the park. It's why Disneyland and Universal Studios feel like their own worlds or even multiple smaller worlds. They feel like their own cities. In fact, when Walt was originally designing Disneyland, he struggled to make the stories and the themes fit. So instead of forcing these themes to align, he did something unique: He created five uniquely different lands.

Let's look at the history of themed entertainment since it touches on things that are central to human nature, which is why they have persisted for hundreds of years and will continue to do so. We'll begin at the beginning and end at the next real beginning: Disneyland.

Buckle up, and keep your hands, arms, and feet inside the ride at all times.

* *

"Oh look, honey! A ride about the history of theme parks!"

"Boring!" I sang.

"All right, well *I* want to ride it," said my mom.

"How could a ride about history possibly be interesting?"

"Well I guess we'll find out, won't we?" she said with a smile.

Beep. Beep. We tapped our Express bands on the sensor and hopped in line. These tiny, lightweight bands had a screen allowing us to access our room keys, credit cards, reservation concierge, and Express passes. As I was fiddling with mine, I heard in the background, "Time travel is not suitable for women who are pregnant or people with any heart conditions, back or neck injuries."

"Did that hologram just say *time travel*?" I asked with rapidly increasing excitement and awe.

"That's what I heard!"

"Cool!"

"See, aren't you glad you decided to ride this with me?" In mom terms, this means, "I told you so." But I was, in fact, glad.

Ten minutes later, we were stepping into our vehicle, which looked more like a glorified Mini Cooper than a time-travel machine. I guess I was expecting Marty McFly's DeLorean— at least his car was cool-looking.

"Step on in, folks. Two people per row! Fasten your seat belts!"

"Remember, everyone, it is imperative that we blend in. So on our next stop, be sure to say as *little* as possible. If you must engage with the locals, please use only a few words. Stray from present-day lingo, and downplay your American accents. American accents won't have developed for quite some years, and it would be quite the scene for someone to hear such a foreign accent. Okay, that's enough chitchat. Now that everyone is ready and seated, let's begin the tour! Our first stop is Copenhagen." He punched in the numbers as he said them: "1-5-8-3. Keep your hands and arms inside the ride at all times as we travel through time to experience the evolution of theme parks!!"

As he hit the gas, we jolted off!! It was mere seconds before we were slowing down again, and I began to see trees around us. We parked the machine in the woods, and once we got out, our conductor went to the trunk of the car and pulled out costumes for us: long dresses and corsets for the ladies, pant suits and coattails for the men.

"Corsets?" I looked at my mom. "Is he kidding?"

"I guess not," she said as he handed her a costume.

After we had all settled into our costumes, the conductor continued, "excellent! Now that we're all suited up..."

I squirmed with discomfort.

"Welcome to Dyrehavsbakken!"

"Bless you," I said.

"Ha! Ha!" He chortled, "No! Dyrehavsbakken, or Bakken for short, is our first amusement park! *Dyrehavsbakken* in Danish roughly translates to 'deer park on a hill.'"

"Catchy," I mumbled.

"People flock here to the natural spring that supposedly has healing powers! But because so many of the locals come here, entertainers of the time saw an opportunity to make a living off those waiting for their water. According to the legend, one of these vendors would make bowls that he claimed would enhance the healing powers of the water. However, if some were chipped or broken, he would allow the children to throw rocks at the bowls to break them, for a small fee. So here is some currency for everyone to use. Take two, pass them around. If you find this man, give him these and you can play the game! Okay, everyone, meet back here in 20 minutes! Have fun!"

Five minutes of walking later, we arrived at Bakken. It was beautiful, breathtaking in fact. As we walked through the trees, into a clearing, the sun sparkled on the water of the spring. It was like a scene from *Bambi*. There were wild deer everywhere, who also were attracted to the clean, crisp water, birds chirping, and not a cloud in the sky. My mom and I both looked at each other, and we were thinking the same thing: "I wish I brought my camera."

"All right," she whispered to me, "let's go find this man with the bowls."

As we strolled about the scene, there were people everywhere waiting in line to get their fresh water. Some had buckets, others had small cups, and others had these beautifully painted bowls. These must be the legendary bowls. Eventually, we stumbled upon a group of locals crowding around something. *CRASH!*

"This must be it!" I said.

We got closer, and there he was: a little old man sitting cross-legged on the ground, painting bowls while also cheering on the small boy who was trying to hit the bowl in front of him with a small rock. I handed the man one of my coins, and he smiled as he set up a bowl for me on a small, wooden pedestal. I picked up a rock off the floor. I closed one eye, aiming at the small bowl. Then I took a deep breath and threw the rock. *CRASH!*

"Yes!!! First try!! Nailed it!!" I yelled.

Oops. I covered my mouth. Everyone was staring at me. This was definitely not how people acted or sounded during this time period. My mother smiled awkwardly as she shuffled me away, heading in the direction of the machine. The conductor was already busy gathering everyone else up. We were busted. All of the locals were very confused and saying things in Danish that I didn't understand. We had to hurry before they started asking questions, or worse, followed us to our car!!

Phew, we made it.

"Hurry, everyone! Buckle up, quickly now! We must be off."

"I'm really sorry about that. It just came out!" I apologized.

"It's really no problem, dear. I just don't want to upset anything in time. That would be bad for the tour group behind us," he joked, trying to make me feel better. "All right, we're all set." He punched in the numbers for our next destination, and we were off again.

"So I hope everyone enjoyed the first amusement park! Bakken certainly is beautiful. However, starting in 1670, the spring was fenced off in order to keep the deer in since the spring was considered part of the royal hunting grounds. Only in 1756, when Frederick V opened it to the public, did Bakken thrive again as a local amusement park. In fact, they got their first roller coaster in 1935, and it is still operational today. Interestingly, Bakken presently is comprised of 160 small, local vendors who operate their own storefronts for everything from the food to the pubs to the merchandise, so even 435 years later, Bakken is still a local treasure.[1] Actually, Bakken isn't all that different than the pleasure gardens of the seventeenth century, which brings us to our next stop, Vauxhall Gardens in 1733!"

He brought the car to a stop and turned to face us. "Vauxhall Gardens was built during the Age of the Gentile, meaning proper manners were valued above all, which includes the way one speaks," the conductor said with his nose held high. "So that means it's best not to engage with the locals and to just stroll about the gardens, listening to the music and enjoying the views. Here is your fee for admission. Meet back at the car in 20 minutes!"[2]

It was pretty, but after about 10 minutes of walking around, I was bored.

"Mom, can we go back to the car? I'm bored. How does this even count as an amusement park?"

"Well think about it. It's an outdoor, enclosed area that charges admission prior to entry and provides entertainment for those who choose to enter. It was a space for people to come, relax, and enjoy themselves among good company."

"Oh, I guess I never thought about it that way."

"Even when you're walking through Disneyland, there are potted plants, hedges, and trees, all made to look beautiful."

She continued on about the flowers, and my mind began to wander: "I wonder if Walt Disney ever looked at these gardens for inspiration? Did he think about the leisurely stroll we're having now and think that he wanted it to be part of his experience?"

"Oh! Maddie!" I was disrupted from my daydream

"We have to go!! We're going to be late!!" my mom said as she pulled me back toward the direction of the time machine. We hurried back to the car as fast as we could, but those dresses didn't provide much movement. Luckily, we made it just in time.

"All right, everyone, you know the drill. Hop back in, fasten seat belts, and let's head off to our next adventure! Did everyone enjoy their time at Vauxhall Gardens?" he asked as he punched in our next destination.

"I liked the music," said the kid in front of me.

"I did too," agreed the conductor with a nod of his head. "All right, our next stop is the 1893 World Columbian Exposition in Chicago! Has anyone ever heard of it before?"

Everybody shook their heads.

"Excellent! First-timers!! Well here we are!" As he said that, the car came to a jolting stop.

"This is a very exciting time for the world of theme parks. The Chicago World's Fair is an example of the first use of themed entertainment! We will be visiting the Midway section of the fair, which acted as the amusement hub!"

We changed our costumes, to equally long and heavy dresses, but at least I didn't have to wear a corset anymore.

As we began walking to our destination, I asked, "I don't get what this is all about? Is this like a street fair or something?"

"Or something... this world fair was designed to commemorate Columbus's discovery of America, and like other world fairs at this time, it was designed to champion American products, arts, feats in technology, and even values and politics. The Eiffel Tower was actually built as part of the Parisian World Fair, just to give you an idea of the scale of these events. In fact, this..." He pointed to the giant wheel in front of us as he continued, "is the first Ferris wheel ever designed!"

As we were walking, I was looking up, admiring the Ferris wheel, and I accidentally bumped into a man carrying rolls of paper and books. "Pardon me, miss," he said.

I nodded, so as not to make the same mistake I'd made in Bakken, but I happened to catch a glimpse of his papers, which were all labeled Burnham and Sullivan. "Huh, nice guy," I thought. "Wonder what he's in a rush for."

Our guide continued, "Here we are in the Midway! The fair opened in 1893, or this year, I guess," he said with a satisfied smile, "which is one of the defining points of themed entertainment."

Mmmrawwhhhh.

"Mom, what was that? Was that a … camel??"

"Look!! Over there!! What is that?" someone else asked.

"Yes, perfect, bringing us to the next section of our tour … welcome to Cairo, everyone!"

"Wait, did we get back in the car or something? What's going on?" I asked.

"As part of the exhibition, Frederic Ward Putnam, who was in charge of the anthropology section of the fair, envisioned bringing the streets of Cairo to Chicago! So you'll see live camels, authentic street vendors, and even a reconstruction of the Luxor Temple! Okay, let's meet back here in an hour. Go explore and have fun! But remember to blend in!!"

"Mom!! Hurry!! I want to ride that camel!!" I said as I pulled her hand.

"Hold on a second! Let's grab a mocha and look at the handicrafts."

"Look! A mummy museum! Do you think they have real live mummies in there? Cool!!"

"No, honey, I'm sure they're just wax reproductions," she hoped.

"Lame..."

The Streets of Cairo at the Chicago World Fair is an example of the first themed, enclosed, immersive land, where the food, the drinks, and the culture were all portrayed as if you were actually in Egypt. As much as this was designed to educate the people of Chicago about the cultures and local ways of Cairo, it was also entertainment because of things like the camel rides, and other various forms of entertainment, like belly-dancing shows. Furthermore, Cairo represented, to the American people, an idealized version of the city, complete with wonder, culture, and amusement, without all of the day-to-day worries of actual city life. Although this particular recreation of Cairo was not exactly an accurate representation, it is still an example of an immersive space for entertainment, the first of its kind.[3]

"Four, five, six, seven, eight. Okay, great, that's everyone! Let's head back to the car now, people. We have a lot more to see before the end of our tour!"

The ride conductor pulled on my seat belt again to check that it was buckled, took his seat, and with the push of a button, stepped on the gas and drove us through time.

As the car slowed, I was jolted forward by the momentum of the stop. My mom instinctively put her arm in front of me.

"Mom, are you going to do that every time? It is a ride, you know?"

She smiled in apology.

"Okay, folks, here we are!! Everybody out, we only have a little time here because we have two more stops after this. The year, 1899, four years after the opening of Sea Lion Park in…Coney Island, New York! While you're here, be sure to check out the first carousel ever created! It is said that the artisan who built the carousel, Charles Looff…"

I snickered under my breath and laughed with a smile, "Looff."

"Actually carved the animals himself! Coney Island is also home to one of the first roller coasters ever, Switchback Railway! All right, enough chitchat, let's head off! Be back in 45 minutes. You don't want to get left behind!"

"Mom! Where should we start?" I asked. "The sea lion show? The waterslide? I think I'm too old for the carousel, so maybe we should skip that. Oh, what about—"

"Okay, take a breath!" She smiled. "Let's just start by walking toward the carousel, because I bet it is really pretty, and then we can decide from there."

We walked along the wooden boardwalk, through the bells and jingles of the carnival games. I was bobbing and weaving through the crowd, trying not to get my eye poked out by all these ladies' hats. "Fashion was weird in this time. These hats could house a family of four underneath them," I thought.

Breaking my daydream again came my mom's voice, saying, "Oh, look! Switchback Railway!"

"Can we go on it, Mom?" As I said that, she was already heading into the line. I smiled.

"Are you coming?"

"Duh!" I exclaimed as I ran to join her in the line.

It was finally our turn. We were ushered to the seats along with four other people.

I whispered to my mom, "Uh, where are the seat belts?"

"I guess that wasn't invented yet," she said in a hushed but slightly worried tone.

"I beg your pardon, sir, but is this quite safe?" she asked the conductor politely.

"So far, miss" he said.

Gulp.

The cart took off at a brisk, but definitely not NASCAR, speed. Up we went over tiny hills, climbing and climbing until we reached the top. I saw the people behind us getting off and running to the other side of the track.

"What is going on?!" I asked.

"Well do hurry up! We are moving to the other cart now," said the lady behind me.

Directly across from the top of the ride was another cart that apparently brought us back down. So my mom and I ran along to the next cart, following everyone else's lead, and before I'd gotten settled into my seat, another conductor was pushing us down the railway. One, two, three hills and the ride was over.

"Well that was wild," I said to my mom, who was also still reeling from the no-seat-belt thing. She nodded her head in agreement.

Coney Island was really the start of what we currently think of as amusement parks. It was, and still is, an enclosed area that charges admission and tickets for individual rides within the parks, which was a new idea—not to mention the fact that it had different parks within the entire Coney Island. After Sea Lion Park, Luna Park, Steeplechase Park, and Dreamworld were added. Its proximity to New York made Coney Island an instant success, with attendance in 1910 sometimes reaching one million people a day.[4] The idea of having multiple parks in one larger area was certainly appealing to Walt as he began his initial design for Disneyland.

"Our next park should be a bit more familiar to all of us, as it's fairly recent, opening in the Roaring Twenties!! However, it wasn't yet what we know it as today. The Knott family started by selling berries, jams, and pies along the State Route 39. Only in 1934, when Mrs. Knott opened her famous Mrs. Knott's Chicken Dinner Restaurant, did tourists begin to flock to this destination. To house and encourage tourists to return, Mr. Knott built several shops and other attractions to entertain the guests while they were waiting for a seat at the restaurant. In 1940, Mr. Knott got the idea to build a replica ghost town on the property, which is where Knott's Berry Farm began its journey!"[5]

"Knott's Berry Farm!! Mom, I've always wanted to go here!"

"Remember, it's not quite the park we know of," cautioned the conductor. "It's still just in the opening years of the Ghost Town attraction. But let's go check it out!"

While we waited for a table at Mrs. Knott's restaurant, we explored the replica gold mine and watched the volcano explode! I was a little unclear on what a miniature volcano was doing in the middle of the Old West, but it was still pretty cool. Plus, Cornelia makes pretty bomb fried chicken. She even came to our table to see how we were enjoying the food!

Back to the time traveler. We sat down; placed all our prizes from Coney Island, our trinkets from Cairo, and our leftover dinner from Knott's under our seats; fastened our seat belts; and blasted back to the present.

On the ride back, I started thinking about how much the ghost town of Knott's Berry Farm reminded me of Frontierland in Disneyland. Maybe that's what Walt was going for. It was his vision to be able to transport us back in time to another place. Now with new technology, we can actually go back in time, but Walt's vision was more idyllic, getting the result of being in the Wild Wild West without actually having to deal with all the hard stuff. Pretty cool.

I was jolted forward again, but this time it was more uncomfortable because I was still full from all the fried chicken.

"Well, folks, it seems like we're a bit ahead of schedule. Before we return back home, who wants to go to opening day of Disneyland?"

We all cheered and bounced in our seats.

"It's settled. July 17, 1955, here we come! Before the actual opening of Disneyland, Walt had been marketing the theme park on television. Every week, he did a segment on his show about different sections of the park. So one night might be about Tomorrowland or Fantasyland. But his technique was so effective that even though 6,000 authentic tickets had been mailed and received, on July 17, 28,000 ticket holders waited in line to get the chance to see the Magic Kingdom."

As our car slowed to a stop, we got out, and immediately I felt the heat of the day. I could actually see the heat rising from the asphalt.

"Welcome, folks, to the original Disneyland!"

"It's hot," I complained. "Is there a water fountain anywhere?"

"Actually, due to a plumbers' strike, few water fountains are functional at the park today. In fact, despite people's enjoyment of Disneyland on opening day, the day itself was a bit of a disaster."

"Really?" asked the little boy next to me.

"Because the opening day came sooner than Walt was anticipating, he and the team stayed the entire night, putting the finishing touches on the park, including pouring the asphalt.

As a result of the heat and the relative newness of the actual asphalt, it was trapping high heels because it was melting. Not to mention, Mr. Toad's Wild Ride shut down due to overcapacity."

"Sounds like a train wreck," I said to my mom.

But despite the lack of amenities and the heat, being there on opening day was special. After going back and seeing all the past examples of amusement and themed entertainment, I understood that standing on Main Street was a realization of Walt Disney's dreams. Walking through the center of Disneyland near his statue was like walking through the pleasure gardens. Being in Fantasyland or Tomorrowland was like being in Cairo at the World Fair. Frontierland was like being back at the Knott's house. Walt combined all the elements of these first pioneers to create a—for back then—one-in-a-million experience.

"Hello? Earth to Maddie!"

"Yeah, sorry." I shook my head.

"Well let's go ride something!"

"Right!!"

"Since Mr. Toad's is down, how about Autopia?"

"Cool! Sounds great!"

*   *

Disneyland is something that will never be finished. It's something that I can keep developing. It will be a live, breathing thing that will need change. A picture is a thing, once you wrap it up and turn it over to Technicolor, you're through. Snow White is a dead issue with me. But I can change the park, because it's alive.

—WALT DISNEY

Walt, after making *Snow White*, felt that he could never make anything as good. People at the premier of the movie cried and gasped, which a cartoon had never been able to accomplish before. Walt Disney knew he had made something incredible. However, due to money troubles and issues at the company, Walt did not have the resources he needed to make another success like *Snow White*.[6] He began to feel like that was the peak of his career, until Disneyland. Disneyland gave him the creative opportunity to imagine a new way of life, not just a picture but something tangible. Similar to Cairo in the World Fair, people could interact with the way of life instead of seeing it in a picture. Walt Disney's idealized version of life, where everything was "clean and colorful," was cemented in the pavement of Main Street, USA.[7] This kind of immersion into a way of life will become the norm for the future of theme parks, and it will only get bigger and better from here.

CHAPTER 3

TOMORROWLAND, 1955

———

"Your birthday is coming up," my brother Jordan said at dinner. "So any ideas of what you want to do for your big 2-0?"

I sat back in my chair and put my fork and knife down to ponder this question. Most people my age would probably have said that they just wanted to go out with their friends to a bar or a party. But the first thing that jumped into my head was probably unusual—so unusual that I had to pause to make sure my family wouldn't laugh.

"I want to go to Disneyland and stay in the Dream Suite."

My father was the first to reply. "Disneyland?" he said, shaking his head and staring at my brother with his head half-cocked. "When are you going to grow up?"

"Hopefully never."

* *

If Frontierland represents our past, Tomorrowland represents our hopeful future.

The original opening of Tomorrowland in 1955 featured rockets flying about, cars zooming by, and a futuristic model home made completely out of plastics.[1] In Walt's dedication of Tomorrowland, he said it was "a vista into a world of wondrous ideas, signifying Man's achievements...a step into the future, with predictions of constructed things to come. Tomorrow offers new frontiers in science, adventure, and ideals. The Atomic Age, the challenge of Outer Space, and the hope for a peaceful, unified world." Whereas sometimes, we can get bogged down in all the burdens of today, Walt wanted us to escape that and look forward to a better tomorrow, one that was fun, peaceful, modern, and safe. Epcot had the same purpose for Walt, just on a bigger scale. "Walt had one foot in the past and one foot in the future."[2]

* *

What Walt was trying to do was "on some level create an image of America that people would like to think exists," a sort of perfection unattainable in everyday life[3]. He wanted people to be able to enter and live in the movies and TV shows that he had created, the happy, clean, safe, brightly colored, perfect worlds that existed inside the safe walls of Disneyland. So even when the economy is down, theme parks are up. People are always searching for the escape from the burdens of reality.

And what's better than real?

According to Walt Disney? *Disneyland.*

I have to agree with Walt on this one.

The reason I love going to Disneyland—more so now than when I was a kid, I think—is because I can be a kid again. Running around, riding rides, getting my face painted, and eating junk food allows me to access a part of my childhood that gets lost in the day-to-day life of work, homework, and society. I can escape. I can be free. I can be whoever I want to be. Maybe Walt really was onto something, because I feel a sense of power over my life when I'm at Disneyland. I can control exactly how my day will go, and at the end of it, I know I will be exhausted but happier than when I came through the turnstile.

One study proves that tourists, after leaving Happy Valley theme park in Shanghai, were happier than when they came in. The 645 tourists were interviewed and asked about their experience upon exiting the park. This study tests whether the CAT (critical appraisal template) appraisal dimensions, which are a series of specific interview questions relating to their feelings toward the experience, point to the signs of delight in a theme-park setting. According to the study, signs of delight can be measured based on their ratings of how much they enjoyed their experience, whether they would come back, how the park fit with their initial expectations, etc.

According to the study's calculations, their hypothesis was confirmed. The guests were delighted by the experience because there were elements of unexpectedness and because their expectations were met or even exceeded.[4]

So, going to theme parks can actually make us happier.

*   *

However, in a world where we've become immersed in technology, why do we crave theme parks that offer us a look at *more technology?*

It's not that we necessarily want more technology, but we do want to see how the future looks with new technology.

We long to see a future version of ourselves in all aspects of our lives. For example, DNA tests like 23andMe are popping up everywhere so participants can get a glimpse of possible future ailments and health. These tests are a scientific crystal ball into the uncertain future of our lives.

Similarly, theme parks have become a place for us to admire the future. While, yes, you can see technology in our daily lives, theme parks—like Disney's Tomorrowland—offer us a way to *experience* the future of technology. And as technology advances, so too do our expectations. In Disney's day, little technology existed, but now you can watch videos—or put on a VR headset—of roller coasters that are more intense than the ones in your hometown, which automatically makes you wish that your theme park had that ride or that you could go visit that ride. When you attend your hometown theme park, you now feel disappointed in the rides that are available to you, because you're thinking about all the other options that exist in the world. Even Disney executive Mike McCullough agrees, people's expectations of theme parks are increasing exponentially.

* *

The theme-park industry is evolving—in many ways because of technology and the impact it has on consumers—as it's done since its origins in the 1500s. As an avid theme-park goer, I am sometimes saddened by this evolution. Every time I went to Disney's California Adventure, I would ride the

Tower of Terror as many times as I could without waiting in line for an hour. So when I found out that my beloved Tower had become a *Guardians of the Galaxy* ride, I was heartbroken.

Disney had kicked out the old favorite and upgraded the wife for the newer, younger, hipper model. I felt betrayed.

But that evolution is part of Walt Disney's original genius. He was never satisfied and always wanted more for his customers. And I learned that the innovative spirit is alive and well. I spoke with the Imagineers and designers who were tasked with innovating on my favorite Tower of Terror. They recognize the importance of change and knew that saying farewell to something beloved by fans like me meant they'd need to do something amazing.

And they have.

Hearing about their process and seeing the models in full scale, I felt like I did while waiting for my wand ceremony—again excited to experience the next evolution, and to see and ride Guardians of the Galaxy—Mission: Breakout!

*   *

I was interviewing Mike McCullough, the environmental design and engineering studio executive for Walt Disney

Imagineering, and we started talking about Disney Shanghai. Since he works in the industry every day, I asked him how he thinks people perceive theme parks today.

"What do you think the customer is thinking about theme parks?"

"That's a big question. I guess over time, because of the world in which we live and how much people are exposed to, I think people expect more. If you look at the original Disneyland, it was really pretty simple. There wasn't anything like it. People didn't have a reference point. And now the bar keeps getting raised and raised."

Back when Disneyland first opened, to raise the bar then, they would announce the opening of a new ride. But today, that's not enough. The opening of an entirely new land is what attracts the customers.

People will certainly want better rides with newer features like virtual-reality headsets. Technology will help solve one of the symptoms of technology, meaning that the parks with the most technologically advanced attractions will see the highest attendance rates. Disney theme parks are still the number-one parks in America, but recently the addition of the Wizarding World of Harry Potter "allowed Universal to replace SeaWorld as the number-two destination in Orlando, behind Disney... The magic of *Harry Potter* conjured a big boost in attendance last year at Universal Studios Hollywood, even as crowds dwindled at Disneyland."[5] Disney representatives have yet to respond to this trend but did note that the company is making big investments, notably Star Wars and Pandora land, that should "pay off in the future. We are investing behind the wealth of great franchises we have in order to deliver magical experiences that exceed our guests' expectations."[6] In addition to Harry Potter World, the Express passes and TapuTapu systems all help Universal to make a play for that number-one spot in the industry, and it seems technology is the means for them to get there. The major theme-park companies are investing huge amounts into new technology. For example, at the last D23 fan event in Anaheim, Disney announced that they will be partnering with Magic Leap, one of the leading companies in augmented reality, with investors like Google and Alibaba. They are creating a newer augmented-reality headset that will feature *Star Wars* games, like holochess.

All of the strides in technology that the theme-park industry is making are only going to make them more desirable for the customers. It does seem odd that in a world where people are looking for off-screen experiences as a chance to come together as human beings, they would be intrigued by the technology theme parks boast. However, it's important to remember that technology is not what makes the theme park. Instead, the technology merely serves to enhance the storyline behind the experience in general. Without the story, there can be no theme park.

* *

"I want to go to Disneyland and stay in the Dream Suite."

Why does an adult woman want nothing more than to stay in the Dream Suite?

Because it's a representation of the power of imagination.

The Dream Suite was intended to be Walt's private apartment for his family and to entertain other VIP guests. Unfortunately, Walt died before his dream could be realized. After his passing, his family set to work on the plans that he had designed and built the suite in New Orleans, above Pirates of the Caribbean. The Dream Suite is the ultimate Disney-nerd paradise. It's loaded with history and magic that make everyday life seem not so mundane. Sitting in front of the fireplace where sparks become fireworks, taking a bath underneath the twinkling stars, and *GONG!!!* goes the grandfather clock that lights up with images of Cinderella at the ball.

The Dream Suite is Walt's vision for life itself. He brought magic to life in that apartment. Sometimes that magic gets made with cartoon-inspired clocks or a fireplace made of fireworks; other times, it comes from wearing a VR headset in a Hogwarts castle or 3D goggles while flying the *Millennium Falcon*—magic nonetheless.

To me, the Dream Suite represents the magic of childlike wonder, and after the family dinner, I had doubled down on the idea from the night before. "I'm serious," I told them both. "If you want to get me anything, it's the Dream Suite. Seriously."

The idea of living for the weekend in such a magic place brought tears to my eyes, and still does.

Now lest you think my brother and father are the master gift givers, they failed, and I still haven't gotten to stay in the Dream Suite. (So, Disney executives reading this book, how about you hook an author up!)

CHAPTER 4

TOMORROWLAND, 2020

——

Beep. Beep. Beep. Beep.

SLAM!

My mom groaned and turned over in her bed after smacking off her iPhone alarm clock. Me? I sprang out of bed like a daisy in spring. I hopped over to my mom's bed and, jumping up and down, yelled, "Good morning! Good morning! It's time! It's time! First day in Tomorrowland!"

"Maddie, give me a minute. I need coffee. Can you order some room service, please?"

"Okay, okay," I said, still brimming with excitement.

After breakfast, we grabbed our backpacks and wristbands, and headed out the door for the day.

Just as the door to our hotel room was about to slam shut, my mom turned to me, panic in her eyes, and asked, "Maddie! Do we have our keys!!"

"Mom, it's 2020, our room keys are all right here," I said as I held up my wristbands that contained our room keys, park tickets, and credit cards.

"Right, right, okay, no need to make me feel old. Let's go."

As we walked through the lobby of our hotel, there were so many other families just like mine who were excited with the possibilities of what today could bring. I passed a girl waving her lightsaber while her mom talked to the concierge, and another boy pulling on the hand of his dad, trying to make him walk faster.

Outside our hotel lobby, we waited for the shuttle that would take us to the park. After about five minutes of bouncing on my heels, our bus finally pulled into the driveway. There was no driver, just a conductor in the passenger seat. The wheel turned itself as it navigated the circular driveway of the hotel and put itself in park, and the door opened by itself to let the passengers load.

"Good morning, sir," I said to the conductor, who was surely there in case of a need for human intervention.

"Good morning, miss, be careful where you point that thing, okay?" He was referring to my wand, and he was right: I needed to put it away until I was safely in the park. I slid my wand into my school robe, just like Harry does in the movies, storing it safely until I was no longer in the presence of muggles.

On our ride to the park, they played a video telling us how to use the glasses that we would get upon entering. After the ticket checkpoint, we would be escorted to a room to receive and test out our glasses. They were small, thin, silver glasses that would act as our park guide for the day, with unlimited possibilities, equipped with our own personal tour guide, Alice. All we had to do was say her name and ask a question, and she would be there to help us make reservations or get directions and even spell locations in Harry Potter Land! The video also showed us sneak peeks of what we would be seeing, like what the Star Wars section of the park looked like, but I was only interested in the world of wizards.

Anticipation was building, almost boiling over.

The bus wheel turned on its own again and slowly came to a stop.

"Finally," I thought. We all stood up and waited for the row in front of us to exit. As I was stepping off the bus steps, my imagination was running wild about what today would be like. Would I get to meet Harry or Hermione? Maybe I could even learn how to play Quidditch!

As my mom and I walked toward the ticket check, we held our wrists up to the sensor, and *beep*, we were through! But before we could really enter the park, we were escorted with 10 other guests to another room, where we received our glasses. They were lightweight, small, and pretty normal looking.

"Back in my day, the virtual-reality glasses were so bulky and heavy, you even had to wear your own backpack!" my mom whispered.

"Okay, folks, you're all set to go! If you have any other questions about anything, even your headsets, please just ask Alice!"

"Who's Alice?" my mom whispered to me.

"Mom, were you not paying attention? She's our glasses' Siri."

"Oh right, okay, well... hey, SIRI, what part of the park is least busy right now?"

I rolled my eyes.

"Why isn't she responding? HEY, SIRI!"

"Mom, she's not deaf. You're just not calling her by the right name."

"Oh yes! Right...hey, Alice, what part of the park is least busy right now?"

"Harry Potter Land is least busy. The wait time for the Hogwarts Tour is five minutes. Would you like to see directions to Harry Potter Land?"

"YES!!!!" I erupted. Best news I'd heard all day.

"Okay, showing directions to Harry Potter Land."

"Well, I guess I didn't need directions, but okay." My mom chuckled.

I laughed too. Alice thought my excitement meant I really wanted those directions, not just being excited that the park I wanted to visit most was least busy.

Regardless, my glasses showed me a larger image of the entire park map. It then charted our course in red, to get from our current location to the entrance of Harry Potter Land. In front of me, I saw a giant red arrow directing us where to walk. It was so cool.

All around me, the surroundings were so full and vibrant. Nothing was left to the imagination. On our way to my destiny, I was walking through a foreign land, complete with street markets and even people speaking in different languages! The detail was impeccable.

After we followed Alice's directions, we arrived at a row of phone booths. Both my mom and I looked at each other, and then around us, in confusion. Then we saw people going into the phone booths but not coming back out again.

I figured it out.

"It's a portal!!! C'mon!!"

Mom and I both hopped into the same phone booth. I searched around for a lever or something to transport us to our next world.

My mind was running, trying to solve the puzzle. "Alice, what do we do once we're in the phone booth?"

"All you have to do is make a phone call."

I lifted the telephone off its cradle nervously.

Beeeeeep.

I looked down at the open phonebook and scrolled my eyes over the list. Leaf blower, leak repairs…and then I saw the name Leavie Muggle and thought, "Hmm, weird name." I realized that it was also in a slightly different font than the rest of the entries on the list, and I knew. That had to be the way to get through the portal! I dialed the number for Leavie Muggle, and sure enough, the ground below us started to move, like we were in an elevator. We descended down below the ground, and all I could see was red brick outside the walls of the telephone booth. The next thing I knew, we were being raised back up into the phone booth, but not the same phone booth. My mom pushed open the door, and I saw wizards, and butter beer, and cauldrons stirring on their own, and the green flash of floo powder. We had made it to Harry Potter Land.

"Okay, that was cool," my mom said, and I could only manage a nod as my eyes and brain adjusted to the world around me. I was paralyzed with awe and excitement. I wasn't sure what to do first.

Almost as if my mom read my mind, she said, "Why don't we take a walk around and see what's what before we make any decisions."

I nodded.

As I began to walk around this new world, I became more comfortable with the magic around me. Eventually, I was able to pull out my wand and perform spells.

"Hey, Alice, show me spell locations please."

"Okay, here are the locations of spells within 20 feet of you."

The map of Harry Potter Land appeared in my glasses. All of the spell locations were red, like they had a thumb tack in them. I located myself on the map and then saw that I had a spell right in front of me. Suddenly, I noticed a spider crawling out of a hole in the wall—it was a gross, big, hairy tarantula. I pointed my wand at it and said, "Arania Exumai!" My wand flashed blue and expelled the spider back into its hole.

"Well done!" said a man's voice behind me.

I had to turn around to see who it was. Since it was just me and my mom, I wasn't expecting a man's voice to interact with us. I turned to see Mr. Weasley clapping and congratulating me. I was in shock. He put his hand out to shake it, and I nervously raised my hand. I wasn't sure if he was a digital projection from the virtual- and augmented-reality glasses or if he was some dude in a suit. But to my surprise, he had a firm grip as we shook hands, and he continued to praise my wizarding skills.

"Ron couldn't get that spell for the longest time! Was that your first go at it?"

"Um...yes, sir, Mr. Weasley, sir."

"Well excellent job! Maybe if you see Ron, you could give him some pointers. Ha Ha!"

"Yes of course, sir, nice to meet you, sir."

"And you! Good luck at Hogwarts this year!"

I looked at my mom with eyes the size of the moon! My jaw was on the floor.

She nodded at the interaction that just took place and smiled at my reaction.

"A little starstruck, Madz?"

"A little," I smiled. "But I'm still confused. How did that just happen?"

"AI robotics," my mom said confidently.

"I'm sorry, what?" It's rare that she knows more than I do about technology.

"Artificial-intelligence robotics, he was a robot."

"Whoa, he seemed so real! I wonder if they have one of Hermoine! She's my hero." As my mind began to plan what I should say if I saw Hermione, my eye caught the school storefront.

"Oh! Mom! The school shop! Can we go in? Pleaasseeee!!"

"Okay," she said reluctantly, because she knew that if she didn't say yes, I would just keep pestering her. Plus, there were so many things I needed.

I walked up to the checkout stand with my notebook, quill, and scarf, where an elf was ringing out customers.

"Are you paying with muggle money?" she croaked.

"Um, yes, ma'am," my mom said, surprised.

I gave her a look that said, "I told you interacting with the robots is crazy cool!"

"One hundred muggle moneys, please. Do you want this taken back to your hotel room?"

"Oh yes, please."

"Room number?"

"423, miss."

"Okay, thank you for shopping with us. Come back again soon."

As she thanked us for our business, she threw the box that contained my items onto a little car that was a little bigger than a toaster. It zoomed off by itself without a driver, no doubt to deliver other customers' items. I watched the little vehicle zoom off through the crowd, navigating legs and buildings. It was such a strange feeling, somehow caught in the middle of fiction and reality—past, present, and future. All this amazing technology was making my dreams of being a wizard come true.

"You hungry?" my mom asked, jolting me out of my daydream.

"Starved," I said.

"All right, Alice, show us spots for lunch."

My mouth was already watering at the thought of butter beer. I wasn't sure how this day could get any better.

PART 2

INTRODUCTION

———

If you've heard of Disneyland, then you've likely heard of Epcot, or the Experimental Prototype Community of Tomorrow. As a kid, I never thought Epcot was worth seeing. I assumed the entire thing was just a boring museum walk of cities all over the world.

Even until recently, I assumed Epcot was just designed for those who didn't want to go on rides. But when I started researching the history of Epcot and Walt's vision for this small city, I was amazed. Not a lot of people know, but Walt was interested in technology not just inside theme parks but also outside of them. In fact, when I interviewed Mike McCullough, who worked at WED Enterprises (Walter Elias Disney Enterprises, which was founded in 1952 and would eventually become what we all know as Walt Disney

Imagineering), he told me that there were teams on the Epcot project that were tasked with the research of other forms of people movers to get people to and from the airport. Walt's visions for the future of technology translated to his vision for Epcot. He hoped Epcot would solve the problems of cities, like congestion, pollution, and even crime. He thought that if he could highlight the problems of today, and present ways to solve them with technology that didn't even exist yet, then it would spur American industry to come up with better solutions to the status quo. Walt envisioned a city that people could actually live and work in, all attached by monorail, from the airport to the city and to the park itself.

It will be a community of tomorrow that will never be complete but will always be introducing, testing, and demonstrating new materials and systems. And Epcot will always be a showcase to the world of ingenuity and imagination of American free enterprise.

—WALT DISNEY

It was a massive-scale project that made WED Enterprises go from 24 to 24,000 people, and one Imagineer equated it to "Disney's space program." Walt Disney was pushing the limits of what he accomplished at Disneyland while also trying to achieve a greater purpose. Through Epcot, Disney was trying to mend the social and political issues that divide our world. When Walt visited the 1960 Winter Olympics, he looked out at the audience, full of Olympic athletes from all over the world, and smiled. He said, "Isn't that amazing? Here are all of these people, different in so many ways, yet united by their hopes and goals and dreams. This is how the world should be." This mindset of peace and harmony with mankind is what inspired him every day to pursue greatness, and to pursue a perfect world.

Here we are, in 2018, trying to improve upon the theme-park experience with the latest technology. Theme-park enthusiasts and employees alike are making the new Epcot, solving the current problems that people have with theme-park worlds with technology that doesn't even exist yet. However, for Walt, it all starts with a good story. If you don't have the back-story to complement the technology, then the park will not be successful. Walt's original philosophy was to "leave the outside world and spend time together with loved ones," and that hasn't changed.

Walt didn't "presume to know all the answers" back then, and I don't presume to know all the answers now. Technology is fleeting because it's always changing and improving, but it's also ever-present and increasingly more ingrained in our lives. How we use technology to enhance our lives will be explored in this section of the book. I want to talk about the latest and greatest in technologies, how they fit into the theme-park world, and also their implications if not used properly in themed entertainment. You'll hear examples of virtual reality being a success and also a complete failure, and why. You'll read stories that seem to come from a sci-fi novel but could be a reality in the next five years. With Walt's vision of the world in mind, theme parks should be aspiring to make the world a better place and bring joy to those who don't have it every day. These next sections discuss how technology can help us achieve this goal.

In the book *How to Be Like Walt*, the authors, Pat Williams and Jim Denney, detail Walt's path for success. First begin by recognizing that change is inevitable. Today we are in a time when change is happening every day, particularly in technology. In order to seize the future as Walt did, we must embrace the change and identify trends and ideas that produce beneficial change. That is what I plan to do in these next chapters.

CHAPTER 5

VIRTUAL REALITY

—

"Okay, folks, in you go! Head straight to the back of the room. Four, Five, Six, Seven, Eight. Oh sorry, miss."

I looked behind me at the woman who was number nine.

"Only eight per ride," the attendant said in a tone that she'd obviously used four hundred times today already.

Once we were all inside, the conductor gestured to the front of the room. A hologram of Eddie Murphy appeared.

"Hey, everybody!! Great place, isn't it? It's a real fixer-upper! But before we go in, everybody take a pair of glasses and an armband from my friend here."

I grabbed both from the attendant and slipped the armband on. Then I moved to the glasses. When I put them on, nothing looked different except for Eddie. He didn't look like a hologram anymore. He looked real, just as real as my dad standing next to me.

"HA HA!! You look hilarious!"

I turned from my dad to look back at Eddie Murphy as a blush colored my cheeks.

"No! Not you, you look great. I'm talking about Sherry over here. She looks crazy in them glasses."

We all couldn't help but crack a little smile at the poor conductor.

*   *

Disney's Haunted Mansion ride opened to guests on August 12, 1969. It's featured at three of the five Magic Kingdoms and to this day remains one of its more popular attractions.

The ride has undergone a series of updates since its debut, including substantial modifications in 1995, 2001, and most recently in 2011.

As reporter Kevin Wong writes, "The Haunted Mansion is one of Disney's greatest creations. It's also an unfixable, thematic mess, which is part of its appeal. It's a mix of ideas, a collection of concepts about how funny or scary to make the best haunted house. It's changed in small ways over the years and is rewarding to appreciate as a kid or an adult."[1]

Part of the charm of the Haunted Mansion is its use over the years as a bit of a "testing ground" for unique concepts to try to make the traditional haunted house into something more "Disney-like" in its scale, its scope, and its appeal. Initially, the Haunted Mansion was thought to be a "museum of the weird," in which the customer would walk through the house and experience weird creatures, illusions, and other oddities. Walt even proposed making the museum into a restaurant concept similar to the Blue Bayou, where you can see the ride happening while you're eating. There was also a discussion about whether to make the ride funny or scary. But eventually, there was a compromise and both Imagineers got their way, with the dark and foreboding atmosphere and fun-spirited ghosts living within. Walt refused to ruin the pristine nature of his park with a run-down mansion. Instead, the beautiful house is creepy within but still has its Disney-esque appeal with the kind inhabitants.

I read a funny story about the manor. One night, the crew working on closing the park for the night forgot to turn off the mechanisms that make the ghosts work, and when the cleaning crew came through the mansion, they were met with quite a surprise: creepy noises and ghosts running around them. The next day, all that was left was a broom in the middle of the foyer. The cleaning crew said they would not be back again.[2]

The Haunted Mansion is a classic Disney ride, and always will be. But it doesn't need to remain the exact same to retain its charm. I believe that with the addition of virtual reality, this classic ride could be repurposed into a new and exciting adventure.

Let's explore what that might be like.

* *

"All right, now that everyone's all situated," holographic Eddie Murphy continued. "Welcome to the Gracey Mansion! I hear y'all are looking to buy it. Excellent! It was built a really long time ago, and it definitely is not cursed. Not at all."

Lightning and thunder struck at that exact moment. Everyone jumped— momentarily startled. He smiled that iconic smile.

"All right, I just need to, um, check on a few things." He looked around nervously. "But, um, enjoy your tour of the mansion.

My pal Ramsey here…" He turned to see that Ramsey wasn't next to him.

"Ramsey, where you at?"

The butler, Ramsey, walked in and stood next to Eddie.

"Here, sir" he said indifferently.

"My man here will take good care of all of you! Bye-bye now!"

As Eddie exited, Ramsey began, "This, in case you were unaware, is the parlour. Here is where we greet the guests, take their coats, etc." He pronounced it "et-set-erah," but he rolled his tongue on the *r*, making him automatically sound fancy.

I looked around me, and it was hard to get past all the cobwebs at first, but underneath all the old, I could see the opulent rug I was standing on and the plush chaise lounge to my left. Even the ceiling was wrapped in gold leaf, and a beautiful painting hung over the mantle.

"Moving on," he said, "do keep up. We don't want anyone getting lost." He said those last words at almost a whisper. The hairs on the back of my neck started to rise.

As we walked through the corridor, I was already creeped out. It was too quiet throughout the house, and I swear the painting's eyes were moving, following me as I walked. Suddenly, I heard faint sounds of music.

"Do you hear music?" I whispered to my dad. As we walked farther down the hallway, Ramsey stopped in his tracks and nervously looked to his left and then to his right, tilting his head as if he too heard something unexpected.

"Please wait here. I just need to check on something. Do not move," he said briskly.

Naturally, however, we all followed behind him. No one wanted to be left alone in this creepy place. He stopped in front of these huge wooden doors, elegantly hand carved. Ramsey pushed on the door handles, and the doors flew open to reveal a ball!

I had to pick my jaw up off the floor. It was like nothing I'd ever seen before. I'd always imagined what it would be like to see a ghost in real life. I thought they'd be see-through like in the movies, or at least all pasty. But they looked just like me, except a little less, well, there—like they were fading away. Ramsey's voice couldn't quite carry over the orchestra, who by the way, were also ghosts.

"What is the meaning of all this! We have guests!!"

It was such a strange feeling. I was scared, but I was also smiling and laughing! Between the sarcastic ghosts and the silly portraits, I was delighted, but things were still jumping out at me and moving when they shouldn't.

After the ride was over, it was almost weird not to be surrounded by ghosts throughout the rest of our day. The transition between reality and what I was seeing in my glasses was so seamless that I forgot what was real and what wasn't.

My dad had ridden this ride before it was virtual reality, and he said I would've been bored to tears if I had ridden that version. Back then people didn't even get to walk around by themselves! This time, I could open any door I wanted, go anywhere I wanted, and interact with whomever I wanted. I even forgot that Ramsey wasn't real! When a knight in the main hall abruptly moved off his podium and scared me half to ghosthood, I moved back to grab Ramsey's arm. Only, of course, it wasn't there! I tend to dislike tour rides because they're like watching a movie but not participating in it, but the virtual reality let me participate however I wanted. It was amazing.

BACKGROUND ON THE TECHNOLOGY

Virtual reality (VR) uses headset technology to transport you to another world. For example, when you put on the headset, you're suddenly transported to the Grand Canyon, looking up at the wall you're about to scale. VR technology is sometimes used in combination with physical objects to create realistic images, sounds, and other sensations that mimic a user's physical presence in the imaginary environment. So to continue with the rock-climbing example, you could use haptic technology (or using touch to communicate with the computer via either a screen or a game controller) to feel as you grab the side of the mountain.

The first references to concepts of virtual reality came from science fiction. Stanley G. Weinbaum's short story "Pygmalion's Spectacles," written in 1935, details a pair of goggles with the ability to record holographs of the character's experiences, including smell and touch.

In 1960, Morton Heilig developed the Sensorama, which he designed to be an on-screen experience encapsulating all five senses in a realistic way.[3]

In 1985, Jaron Lanier, an early VR pioneer, and his company, VPL Research, developed a DataSuit, which is a full-body outfit with sensors for measuring the movement of arms, legs, and the trunk to create a VR experience.[4]

By 2016, there were at least 230 companies developing VR-related products.[5] Google, Amazon, Microsoft, Sony, and Samsung all have dedicated augmented- and virtual-reality groups. Oculus is currently considered the master of the VR world. Oculus Rift's display technology combined with a low-latency tracking system leaves the user feeling like they're really in the virtual space. According to their website, you don't even have to know how to use their hand controllers, because your intuitive actions in the virtual world are as natural as they are in reality.

These are some of the most promising applications for virtual reality:

- Video games
- Cinema and entertainment
- Healthcare and clinical therapies
- Architectural and urban design
- Music and concerts

Here are some of the most intriguing applications of virtual reality:

- **Experience skydiving without jumping from an airplane.** Reporter Jonathan Vanian was able to test out Sony's virtual-reality system and use its game Social VR. He described the experience, saying, "It was in this cartoony and adorable world where I felt my adrenaline rush after using a virtual beach ball to bounce myself up into the sky. Eventually, I bounced so high that I could see tiny trees below, hilltops, and the vast dark blue ocean across the horizon. But of course, what goes up must come down, and I soon found myself free-falling back to Earth. In reality, my legs were firmly planted on the conference room floor. But seeing the horizon get bigger and closer as I plummeted was enough to cause my stomach to drop and my heart to race. I've never skydived, but I got a little taste of it in virtual reality."[6]

- **Experience getting attacked by giant robots.** RIGS: Mechanized Combat League takes the user to a futuristic arena in Dubai, straps them into a giant robot, and lets them battle it out Transformer style. The objective of the game is a little similar to basketball: get the ball in the hoop while trying not to be killed by the robot opponents. Use laser cannons, jump, run, and dunk in the arena, but try not to get dizzy from checking back to see who is on your tail![7]

- **Meet an 80-foot whale.** Wevr's theBlu transports the user to an underwater sunken ship in the middle of the ocean. One user recalls, "This is the largest creature on the planet, and you can feel what it's like to come face to face with it. The whale's eye is about as big as your head, and the size of the beast is overwhelming. You really start to feel like you're underwater." Although his 10-year-old nephew was unimpressed with the lack of shooting in the experience, his daughter was blown away.[8]

- **Perform surgery, virtually.** Osso VR is a training platform for surgeons, nurses, interns, and students of all levels that allows the user to practice and perform surgery! Imagine having to study for a surgical exam and not being able to actually practice your techniques! Osso revolutionizes the way students will learn to become surgeons. Even a high school student looking to become a doctor can begin practicing early and develop her or his skills.[9]

CURRENT APPLICATIONS IN THEME PARKS

"Everyone in the industry knows that virtual reality and augmented reality will dominate the conversation for the coming years," a former Disney vice president shared with me.

And the reality (see what I did there?) is that there are numerous rides and experiences today that are utilizing elements of virtual and augmented reality.

One such example is Alton Towers' roller coaster *Galactica.*

When you enter this ride and sit in your seat, your feet are dangling below you. You strap on your virtual-reality headsets and headphones. Attendants come by to check that you're all strapped in and your headsets are working properly. Before you pull out of the station, the seat rotates so that you're parallel with the floor, like how Superman flies in the sky.

As the ride begins, you exit a space station and begin flying through space, avoiding collisions with asteroids and satellites. The addition of virtual reality makes the ride surprising in a new way. When you ride a traditional roller coaster, you have a pretty good idea of what's coming because you can see the track. But with a headset, what's right in front of you could change at any moment. The programming combined with the G-force of the rides adds a new dimension to the experience that is surprising and exciting.

Another example of virtual reality being used to enhance a traditional theme park experience is *The Void*.

The Void works to create "hyper-reality" experiences, where the customer goes into a room with their headset on and actually touches and feels their surroundings as they're seeing them. The Void, in comparison to other VR companies like Oculus, boast completely immersive experiences outside of your living room. Not only are you seeing and walking through a virtual space, but you're also interacting with real objects. For example, say you see a staircase ahead of you. During your Void experience, you would actually grab the railing on your right-hand side as you descend.

When you enter The Void experience, you are helped into

your gear, which includes a helmet complete with glasses, headphones, and gesture-recognition technology. Gesture recognition could come about in a multitude of ways, sometimes through game controllers, others through gloves with sensors built in. Either way, both controllers and gloves use sensors to interpret your movements to correspond to what the computer projects and what you are seeing through a variety of computer algorithms. During The Void, you're also asked to carry a portable computer backpack, which looks a lot like the heavy laser-tag backpacks. In fact, that's what it is.

Think of [The Void] as the rainy-Saturday laser tag of the future, but on steroids. Lots of steroids.[10]

The Void recently announced that they are installing a location in Downtown Disney as a *Star Wars*-themed attraction. The customer can step into the world of a stormtrooper and explore a starship in a galaxy far, far away, in a very real way. The Void is probably the most advanced version of virtual reality that we have seen. Taking the virtual experience outside the home, and making it more immersive, makes The Void the first of its kind. This kind of technology has obvious applications to the theme-park industry. Even The Void's first location in Salt Lake has been called, by Vice, the first virtual-reality theme park. The Void is definitely a step in the right direction of where VR could be. However, take off the glasses and you're just in a dark room with lights and sensors. Not to mention that this experience occurs in a confined space and allows for a maximum of four people per group.

Another VR company looking to change the game is Dreamscape. Dreamscape's "vision is to change what VR has been," offered Adam Aron, chief executive of AMC Entertainment, after experiencing the Dreamscape experience. "Away from just a heightened level of video game and toward cinematic storytelling—and we think it's what consumers have been waiting for."[11]

Dreamscape Immersive is run by CEO Bruce Vaughn, who was previously Disney's head of Imagineering. Dreamscape seeks to create a movie theater that allows customers to purchase a ticket and actually step inside the movie, not just watch it.

Dreamscape works by using more than a dozen cameras and sensors around the space to transport you anywhere you want to go, from courtside at the Lakers game to underwater in Atlantis. You put on the VR goggles, a pair of gloves, a portable computer, and the attendant clips a small device to your shoes, and then you're off to your Indiana Jones adventure through an ancient temple. You can even pass objects between you and your friends. So let's say your friend needs the torch light—no problem! Similar to The Void, this experience transports you to a different place and time with the help of virtual reality. The difference between the two is the small sensor Dreamscape uses to approximate your body movements. This helps to make your virtual friends not look weird when they move in real life.[12]

However, even this experience, when it comes out, is still limited to between two and six people.

FUTURE OF VIRTUAL REALITY IN THEME PARKS

Some skeptics have mused that as virtual-reality technology becomes cheaper and better—such that home use becomes the norm for mass-market consumers—consumers won't need or want to go to theme parks.

Color me a skeptic of the skeptics.

Theme parks are always on the bleeding edge of technology, and while certain applications may become used in homes, those are likely the applications or scenarios that theme parks are already making available to consumers *today.*

What we'll see ahead is a mixture of newer, more immersive, and more integrated applications of VR into theme-park rides and attractions.

Chen Jianli, chief executive of the Oriental Science Fiction Valley theme park, is already thinking on this level. He plans to incorporate virtual reality throughout the entire park. You won't actually wear the headsets around the park, but all of the attractions and rides are virtual reality. This mixture of virtual reality and the real park eliminates the moment during The Void, where if you take off your glasses, your environment is no longer interesting. At Oriental Science Fiction Valley, you're able to take off your goggles after your ride and experience the park normally.

The park opened in February 2018, and it will be interesting to see how people respond. It's the first of its kind, and personally, I am interested to know how the customers respond to the story. The reason franchises like Universal are successful is because of the IP associated with them. This particular park doesn't have the same kind of popular characters to relate to, so it will be interesting to see whether people feel that this experience is worth coming back to time and time again.

* *

Today's virtual-reality applications are more limited than a future world imagined by Jianli and the Oriental Science Fiction Valley theme park.

"The dozen pounds of electronics weighing me down," one participant of The Void experience recalled, presents a challenge to customers. Most VR—particularly in high-usage scenarios like theme parks—requires heavy and bulky technology.

Even as a kid, I remember going to laser tag and having to schlep that heavy backpack around. Imagine if you didn't have to wear that bulky tech? Experts in the industry, like Nolan Bushnell, Imagineering employees, and tech experts like Jason Rauhoff, see the future as having a headset-less experience.

Nolan Bushnell thinks that we are at least 30 years from the "wet wear" experience, but it's definitely what the crystal ball is projecting. In the above Haunted Mansion example, the guest is wearing just glasses and an armband. I hope to see the future of virtual reality with a Google Glass-type experience so there is no bulky helmet. Also, instead of the computer, companies like startup Thalmic Labs are creating the armband that can interact with and control electronic devices through gestures, which could hopefully in the future eliminate the issue of the backpack.

According to the industry, the main issue with virtual reality is motion sickness. There have been multiple complaints about virtual-reality roller coasters making the riders sick to their stomach, and not in the way they're supposed to.

Another current limitation to VR in theme parks is the scale. It's one thing to have a singular VR experience at your theme park, but it's another to make it the whole experience. Both The Void and Dreamscape only allow up to six participants at a time. But let's think bigger! Imagine being able to walk around any land of your dreams all day and incorporate food, beverages, and merchandise into it.

We should expect to see large sections of theme parks leverage the cost effectiveness of VR to upgrade older and outdated elements of their rides. Just look at the robots that you see in rides like Pirates of the Caribbean, and imagine the cost of creation and maintenance of the mediocre props. When the entire New Orleans Square in Disneyland was completed in the sixties, it cost $15 million—$8 million (roughly $65 million today) of which was dedicated just to the Pirates ride.[13]

In comparison, the Haunted Mansion, which opened only a few years after Pirates, cost around $46 million in today's terms.[14] That's substantially less than the cost of Pirates, and that's because of the animatronics. Even Mohamed Newera, a retired Imagineer who worked on Disneyland Paris, which also features the pirate ride, commented that robots "are the most expensive things."

Both virtual and augmented reality can help to mitigate some of those costs. Imagine riding Pirates of the Caribbean with a VR headset and actually seeing Johnny Depp sword fighting and swinging from the ropes above you, instead of in static robot form. Well, Disneyland Shanghai actually did. They found that the addition of virtual reality really helped their park because the storyline was much more compelling, and it eliminated the cost of the robots, or audio animatronics, as Disney calls it. When Imagineer Mohamed Newera was working on the Pirates project in Shanghai, he recalled, since robots were expensive, "In Shanghai, they did away with 240 robots, and they kept only 50 robots. Virtual reality replaced the need to build so many robots and still save a lot of money."

Immersive technologies are also excellent for holiday use because the roller coasters can, with a simple reprogramming, make the ride Halloween or holiday themed.

Older rides can be repurposed by adding a new VR experience to them. A ride like California Screamin', a tried-and-true Disneyland California Adventure veteran, could easily be converted to Moana's Wild Wave, without the headache of tearing the ride down and starting from scratch.

* *

Virtual reality is already being implemented into the theme-park industry. More and more companies are investing in virtual and augmented reality as parts of their theme-park experience. Instead of like in the past, when new rides would drive more customers, in the future, the theme park with the latest and greatest technologies will attract the most customers. The difference between the type of virtual reality that will succeed and those that won't is whether they start with a story.

Hub Zero in Dubai is an example of virtual reality not working, according to at least one former Disney Imagineer. In Time Zombie, the customer is transported to a zombie-infested world, and they have to shoot the zombies to survive. By using virtual reality, the user is brought into the video-game screen instead of being on the other side.

However, in one Imagineer's opinion, this ride did not encourage repeat customers, and he felt it was a one-time experience. This particular experience's failure was because "nothing kicks ass harder than a good story." Time Zombie at Dubai's Hub Zero was lacking just that.

A good story that is relatable and nostalgic, and evokes emotion is the other half of why people visit theme parks, the first half being chiefly roller coasters. The future of theme parks is in its "sensitive" relationship with immersive technologies, meaning you need the story before you can add the headset. I believe that virtual reality and other immersive technologies have the potential to make theme parks the best they've ever been in terms of cost effectiveness, immersive-ness, and improving the all-around memorability of the guest experience.

In looking to the future, parks should not start with the technology. Instead, theme park designers must remember that, as Disney said, it all started with a mouse. Start with the story, and use the technology to make the storytelling experience more fun.

While a future of in-home virtual-reality devices certainly could pull market share from theme parks, my expectation is that the power of more robust and bleeding-edge technologies—unavailable in home systems—coupled with the power of stories and brands consumers love will enable virtual reality to play a huge role in the future of the theme park.

CHAPTER 6

AUGMENTED REALITY

———

You already own the world's most amazing computer. You're using it right now, to think with and blink with and maybe even smile with. It's your one-of-a-kind mind. And with it, you can do incredible things. Magic things.

—MAGIC LEAP

I was standing impatiently by the hotel room door, playing with the chain lock, waiting for my dad to be ready. The silver walls were reflecting the lights from the lamps. The galaxy outside our window didn't provide much sunlight.

The clock read 7:00 a.m. on Saturday morning, early opening of the park for guests staying at the Star Wars Hotel in Los Angeles.

"C'mon, Dad, we're going to be late!!" I whined impatiently.

"Okay, okay, I'm coming," he said. I pulled his hand toward the elevator, and as we descended to the lobby, we both put on our augmented-reality glasses and connected via Bluetooth the small "digital ticket" we were each wearing around our wrists. I was bursting with excitement.

This was going to be epic.

I *love Star Wars.*

*   *

While many people talk about virtual reality as if it's *the* most important technological innovation since the smartphone, let's not forget its cousin, augmented reality. Samsung, Dell, Sony, HTC, Oculus (a division of Facebook), and others have all rushed to create virtual-reality headsets to capitalize on the VR wave.

Not on that list: Apple.

"Tim Cook's preference is for augmented reality (AR), a technology for overlaying digital information on objects in the real-world seen via a smartphone camera or headset." *The Economist* reported from their conversation with Tim Cook, Apple's chief executive: "The display technology is still not good enough to provide a satisfactory VR experience. He also believes that virtual reality (even more than a mobile phone) is too isolating and anti-social."

And so Apple has gone deep on augmented reality, with "hundreds" of its engineers focused on the technology. The latest operating system (iOS 11) includes a set of tools known as ARKit for building augmented-reality applications for the iPhone and iPad. Some 1,000 or so AR applications can now be downloaded from the company's App Store.

"Augmented reality amplifies human performance instead of isolating people," Cook tells *The Economist*. "Investing in AR also makes more business sense. Unlike VR, which is essentially a game-playing technology, AR has many potential uses. It is expected to make big impacts in industry, education, training, healthcare and travel, as well as sports and entertainment."[1]

Numerous theme-park experts I've spoken with tend to agree with Cook—VR may seem like the future, but AR is the present and also the future, and even Oculus has shut down their VR Story Studio.

* *

"What do you think it's going to be like?" I asked my dad as I adjusted my glasses. "Do you think it'll look just like the movies?"

"If it's anything close to the reviews, then it's going to be *exactly* like the movies."

The glasses were remarkably low-tech in appearance—nothing like the virtual-reality headsets I'd seen online. They were a simple pair of glasses, perhaps slightly heavier than sunglasses, with a smallish box on the side that held its electronics, including the Bluetooth. The small watch on my wrist, connected to my glasses, contained our tickets and photos, and in my dad's case, the credit cards, which we would use for snacks and merchandise. I'd been dying for my own lightsaber.

The glasses were clear, and I noticed a small blinking light in the upper right corner of my view. Underneath, a cursor was blinking.

Suddenly—almost as if being typed—words flashed under the blinking light.

"Welcome, Commander Madison Kelley."

"Cool," I thought to myself. "I'm a commander. Wonder what my dad is."

Before I could ask him, the elevator door opened and we walked through the lobby of our spaceship. The small text in the corner of my view read, Proceed to the Escape Pods.

Once we were in our pod, the pilot took off and kicked into hyperdrive. Blue and white lights flew by the window like paint pouring out of a can and onto a blank canvas, until the pilot began to slow the shuttle. We could see the planet Naboo. It was covered in green trees and sparkling lakes. My eyes widened with awe and wonder. As our craft began to circle and descend on the landing strip, I could already see C-3PO rushing to greet us. The wall in front of us lowered into a ramp for us to walk down.

"Good morning, sir! How are you today?" he asked my dad.

Okay, now I was *really* freaking out.

"Excellent, C3, and yourself?"

"Oh, just wonderful, sir, thank you for asking. Do you have everything you need for your journey today?"

"I believe so!"

"Remember, if you need anything, please do not hesitate to ask. My main function is human-cyborg relations, but you should be able to access everything through your glasses."

"Thanks, C3!" I responded.

"You're most welcome, Commander Maddie—may I call you Maddie, or would you prefer your given name of Madison? Oh, and Happy Birthday!"

"Maddie is fine, and thank you, C3! See you later!"

"Most certainly, commander."

As we started to walk toward the exit of the pad and also the security and ticket check center, I looked up to see other spacecrafts zooming above me. Aliens of all shapes and sizes were busily bustling about the air strip. The mollusk-looking creature in front of me left the grossest slime trail behind him. I had to stay five feet back just to avoid stepping in the goo.

Squish.

"Oh no..." my dad said as he checked the bottom of his shoe. Yup, he stepped in the slime. As he turned over his foot in his hand, we could both see the green goo dripping from his shoe.

Gross.

* *

The slime? Not real.

Using augmented-reality technology coupled with some potential tricks, like slight vibrations in the floor, could completely trick the human mind and make it appear real to both of us.

C-3PO? Also a product of the glasses and a small speaker embedded into the glasses to make it sound as if he was speaking directly to me—or us. Using techniques like sound echoing, you can make sounds appear to be hundreds of feet or just a few feet away. The camera of the glasses captures the user's depth in relation to her or his environment. This information is then sent through a digital 3D engine that translates the sound within the digital space. The sound is then sent back into the real world using the speaker on your glasses.

The other aliens or even the other crafts in the air?

All projected using augmented-reality technology.

And while this scenario is a product of my imagination—not much reality in this one (yet)—this fantasy will become a reality (or at least a projected image into our reality).

Luckily for the sci-fi and theme-park fans, it is already becoming very real *very* soon.

BACKGROUND ON THE TECHNOLOGY

Augmented reality, with the help of glasses, a headset, or a screen, adds digital elements into the real world. For example, as you walk down the street, large digital arrows point you in the direction you want to go. You're seeing the real street in front of you, but the digital arrow has been added. The computer projects a live view of a physical, real-world environment, and the digital elements are either computer-generated or extracted real-world inputs such as sound, video, graphics, haptics (using touch to control and interact with the computer), or GPS information.

The first reference to anything resembling augmented reality was by an author: L. Frank Baum, best known as the author of *The Wizard of Oz*. In his book *The Master Key*, he mentions the idea of electronic glasses that overlay information onto the real world.

This wasn't the last time augmented reality appeared in popular literature or television. Even the Terminator had augmented-reality spectacles that he used to analyze the truck he stole. In 1992, Louis Rosenberg developed the first functioning AR system, called virtual fixtures, for the US Air Force Research Laboratory. The user would use two robotic arms and a headset to see the robot arms in the place of their arms. The system also showed simulated barriers and fields designed to assist the user while performing real physical tasks. By overlaying these augmented features, they actually proved that the technology could improve the dexterity of human beings. Every single year after 1992 has seen some improvement in the AR industry, and even now it continues to be improved. [2]

When starting the research for this book, I wondered, how different can virtual and augmented reality really be? They both alter reality, but virtual reality implants you in an unreal space, whereas augmented reality implants virtual objects into your real world. Think of virtual reality and augmented reality as cousins. And while many people are interested in virtual reality, there are just as many interested in augmented reality.

These are some of the most prominent applications of augmented reality today:

- Architecture
- Online retailing
- Education
- Emergency response systems
- The military
- Navigation
- Music
- Video games

Here are some intriguing applications of augmented reality:

- **Learn about a zebra by seeing one right in front of you.** Initially I had assumed that augmented reality would be just for fun, but it actually can have real impacts on education. In fact, there already exists a myriad of educational augmented-reality apps, for example, from Bitar Labs. They use mobile augmented reality (augmented reality on your tablet or iPhone) in tandem with their books, which range from space education to animal studies. When using the app, you hold your tablet or phone up to the book, and the zebra you're looking at will become three dimensional in front of you, and it even makes sounds and jumps around.[3]

- **Place your dream sofa in your house before buying it.** The company Augment implants digital commerce items in front of you. So let's say you're online looking to get a new sofa for your house, but you're not sure how it will actually look inside your living room. No problem, Augment allows you to put that couch in the exact spot you need it, in full size.[4]

- **Get a tattoo!** InkHunter is the app that could help you decide whether or not to get that tattoo you've been talking about for years. Their camera-based technology allows any tattoo of your design, or of theirs, to be placed on any part of your body, oriented to your specifications, so you can get an accurate depiction of what your tat would look like without actually getting inked.[5]

- **Find a good place to eat in a new city.** Just open your Wikitude World Browser app, and use the camera to show your browser where you are. The app will then pop up everything from restaurant recommendations based on TripAdvisor to historical information from Wikipedia. Wikitude is considered the "king of all augmented reality browsers, and in a way, serves as a third eye of sorts."[6]

CURRENT APPLICATIONS IN THEME PARKS

Universal Studios is currently the only theme park that has a true augmented-reality experience (as of March 2018). In their Islands of Adventure, specifically at Jurassic Park, customers can interact with dinosaurs through a large screen. They use live rendering to enter you into a space with characters, objects, or animals that you interact with. INDE offers a variety of experiences including large- and small-screen activities, called "mobile AR."

Another front-runner in the augmented-reality game is Magic Leap. They are known in the industry for their mystique factor. They have yet to release a fully developed product yet, but nonetheless boast investors like J.P. Morgan and Google, and even celebrities like Beyonce have been granted a sneak peek at their tech.[7] They seem to be very close to releasing something to the public, since they stated in September that they would be sending out their technology to a few users in six months. What sets them apart from the rest of the augmented community is that, "no company has successfully built [augmented-reality] glasses that people will casually wear in daily life. Magic Leap has shown progress on the very first and very last steps: the core visual technology, and the entertainment content." Their headset is designed using "dynamic digitized light fused signals" that give the digital objects a sense of realness. Basically, their glasses will capture not just parts of the light rays, like in a photograph, but instead the whole ray, including its direction. This allows our eyes, with the help of the technology, to refocus the information from the ray and allow us to see the depth perception of the scene in front of us in 3D. "The Magic Leap headset is said to use a light-field display powered by a novel array of nano-structures, giving the digital imagery true-to-life depth cues which in turn makes the projected image seem more real."[8]

This is a huge step in the right direction, technology wise. There are a million companies that are working with augmented-reality headsets, but to have an experience without the obstructiveness of the heavy glasses, which means even more mobility, is on the right track for the kind of technology that I will be talking about later in this chapter. INDE is also using a headset-less experience, but their technology, to me, is still not as interesting as Magic Leap. With INDE, the user is still separated from the screen. The *virtual* you that is being projected is interacting in real time with the characters, not the real you.

It would be much more powerful to actually be the one physically in the screen. However, the scale of their experience allows you to share the moment with more than a few friends, which is thinking more large scale than other VR experiences.

While there are limited examples of pure AR technology integrations, we can learn a lot about its potential applications by examining how theme parks have used 3D glasses. Originally used primarily in movies, innovators began to look for ways to incorporate 3D projection into more traditional thrill rides.

One of the first to do so was Universal's Amazing Adventures of Spider-Man, Islands of Adventure, which debuted in 1999 at the Florida park. Tamara Hinson of CNN says, "This ride took three years to build, but it appears the effort was worthwhile. It's won several awards and has been awarded Amusement Today's Golden Ticket award for best dark ride for 12 consecutive years."

"One real game changer is the Spider-Man ride system and its multimedia technology," says Maximilian Roeser at MACK Rides. "The 3D effects are so highly detailed and synchronized to the movement of the cars that you hardly can tell what is projection and what is real set."

Comparing rides using these technologies, it's clear that 3D does add to the experience. When I went to Universal and rode the Despicable Me ride, I didn't even bother putting my hand up when things came out at me, because I was in front of a screen. I knew what I was seeing wasn't real.

But when I rode Escape from Gringotts in Harry Potter World, I actually thought there were bits of the wall coming toward my face, and I blocked it with my hand. The difference between those experiences is the reality of my surroundings. I was actually moving through space instead of having the seat beneath me staying stationary like in Despicable Me. In Escape from Gringotts, I had no idea where I was, and that lack of understanding made it more believable for my senses, which is why it was instinct to block my face. 4K projection has begun to render some of these differences moot, but in an attempt to elicit more "hands in front of your face" reactions like Gringotts, designers are seeing AR as a way.

And these types of reactions will only be enhanced when the experience is more customized and realistic for *each* theme-park goer using augmented-reality glasses and screens.

FUTURE OF AUGMENTED REALITY
IN THEME PARKS

Theme parks have long used 3D glasses to bring enhanced experiences to the rides. But many fans have felt the glasses are remnants of prior times and prior technologies.

Robert Niles, a theme-park reporter for the *Orange County Register*, finds 3D glasses underwhelming to say the least.

He says, "When Universal Studios Hollywood opened its Wizarding World of Harry Potter land last year, it offered its Harry Potter and the Forbidden Journey ride in 3D, a change from the ride's original installation in Orlando, where it continues to run in traditional 2D. But earlier this year, Universal Studios Hollywood dumped the 3D version in favor of a new 4K projection system on the Potter ride. I couldn't be happier with the switch. Last month I rode the original in Orlando, and Hollywood's version just blows it away for clarity, brightness, and visual detail. Even without the 3D effects, the 4K projection just makes the entire experience feel more 'real.'" I definitely felt the realness when I rode Escape from Gringotts. I swear Voldemort looked into my soul, and to be completely honest, I cried a little from fear."

The advances in 4K projection that produces hyper-realness that made a grown woman tear up is beginning to make 3D glasses irrelevant. Niles states, "Wearing some of the bulky 3D glasses also narrows your field of vision, defying the supposed benefit of 3D making an image appear more lifelike."

And it's this innovation—4K projection—that opens the door for augmented reality that can offer enhancements that were never possible by simply viewing an image through polarized lenses.

* *

Much like Apple, Disney remains confident in AR's capabilities and has been investing in the technology already.

When his Imagineers brought him a set of ideas for virtual-reality integrations, Bob Iger's reaction was pointed:

"Don't even think about it," he said.

And some consumers already agree, citing concerns about the bulky headsets—especially for children, the cleanliness of sharing VR headsets, and even the issues with motion sickness that many first-time users of the services face. Theme parks want you to save the queasiness for roller-coaster drops, not their imagery projections.

Iger, the CEO of Disney, believes that AR is the next big thing.

At Disney's D23 convention, the company released a teaser to their new augmented-reality helmet that allows the user to play games like holochess, or even train like a Jedi with a Bluetooth-connected lightsaber. The ability of AR to integrate sounds, sights, and even interactions with other sensors and computing systems such as doors, floors, or robots make the technology more than promising.

Augmented reality has the potential to improve theme parks immensely. And while virtual reality certainly can help the parks be more effective, it might be further off than augmented reality, which companies like Apple have shown an ability to rapidly develop applications of.

Augmented reality—like VR—can help theme parks avoid building expensive infrastructure, robotics, and decorative facades on the buildings. In my example at the start of the chapter, the cost savings of an augmented C-3PO vs. a robotic figure is massive, not to mention the ability to continually improve and enhance the characters.

Likewise, augmented reality can help to fill in the blanks for the customer, creating an equally immersive and interesting experience while mitigating the expenses. While the park should build certain infrastructure, like the security and ticket check building, the detailed facades, like those in Harry Potter World, would not have to be as in depth—augmented reality could help fill in the details. Furthermore, AR rides are more immersive and more believable, making them more exciting.

Perhaps the reason that we've only seen one AR experience in theme parks so far is due to the technology. The goggles are still being developed into their best versions, and without the goggles, we are limited to on-screen experiences. I think theme-park executives recognize the power of this technology and are waiting for the right version of AR to come out before they apply it to their park. Universal's Jurassic Park example is only a baby step in the direction that AR is heading, and while Universal gets to boast that they were the first, they will certainly not be the last.

CHAPTER 7

PERSONALIZATION

—

With the rise of data mining and cookies, we're used to having our information used against us. I clicked ONCE on Saks Fifth Avenue's website, and now my Facebook is flooded with outfits that I might like to buy and Instagram suggests profiles that I might like to follow.

These seem like annoying inconveniences, flooding our feeds with information we don't need. But in reality, this technology has changed us. Most websites use our information to understand our patterns and habits to draw conclusions about our preferences in order to provide a more tailored experience. As much as we might hate these pesky advertisements, at the end of the day, we expect things to be tailored to us now. Everything from our iPhones to our Bitmojis is customizable. *Entrepreneur* notes how the most effective way to market to millennials is to do so by making them feel that their business can apply to their personal needs and likes. They cite Primark as a company that allows buyers to upload pictures of themselves to their website, where they can rate and like each other's looks. Primark will even invite people to register for prizes and initiate chats between users about fashion. Primark is appealing to the younger generation's "narcissism" and gives them the "opportunity to develop [their] own unique style profile and to share fashion ideas and inspirations with others."[1]

Since the coming generations will increasingly desire more tailored experiences, it makes sense to apply this thought process to the theme-park experience. Imagine if you could personalize your entire experience, ranging from rides to food, from merchandise to the shows. When I went to Disneyland in high school, I would do something similar, just without technology to aid me! Every time I went, I learned something from the previous experience and would apply it to my next one, just like how cookies learn from our experiences on a website. Each time after a trip to Disneyland, I would look at the order of rides I chose, the food I ate, and the merchandise I had just bought and make conclusions like, "Okay, avoid Main Street during the parade because it's a madhouse trying to get anywhere around that time." If I'd had access to better technology, I would have been able to do my experience right the first time!

Have you ever been to Bill Gates' house? Yeah, me neither, maybe someday. Not only is his house huge, but it also has technology that makes the Crestron system in my house look like TiVo. He has taken the term *smart house* to the next level. Upon entering his home, you are given an electronic pin/camera for facial recognition. "The electronic pin you wear will tell the house who and where you are, and the house will use this information to try to meet and even anticipate your needs—all as unobtrusively as possible."[2] By interacting with the sensors all over the house, the system is able to learn your preferences and adjust accordingly. So if I prefer natural light and Rembrandt, when I walk into the living room, the shades automatically rise and the paintings change from Pollock to Rembrandt in a second. All with the touch of your iPhone, you can adjust your preferences as you go along, so the house will automatically update for next time. What would happen if you applied this to theme parks?

Imagine you have just purchased your hotel and park tickets online to stay at the new theme park, Technoland! You click on your email confirmation and open the PDF that details your reservation: what kind of room, what kind of park tickets, what amenities your hotel provides, etc. Once you've confirmed everything is in order, you click on the next PDF that looks like a survey of some kind. On this page are questions like, what kind of food do you like, what's your favorite junk food, and what rides would you like to check out, coupled with pictures and videos of the rides themselves. You click and type in the appropriate areas, detailing your preferences for your experience.

A few weeks go by, and it's finally time to head to Technoland! You're jittery with excitement in the car on the way from the airport to the hotel. When you arrive, you head straight to the check-in desk and get your wristband, which acts as your room key, express pass, and credit card all over the parks. Along with your band, they show you on your Technoland app an itinerary of sorts, telling you which rides to ride, where to grab food and snacks in between, and where to get your face painted, all while avoiding the lines and crowds. It even has included a character breakfast with your favorite character before the park opens to hotel guests!

The next morning, you wake up and prepare for your character breakfast, complete with all your favorite breakfast items. You're expecting the big felt-costumed actors that never speak, but this park is different. These are artificial-intelligence robots that look and talk just like the Mr. Gru you know and love! After breakfast, you walk through the downtown area of the park, scanning the stores for all the merchandise you want to buy later, until you arrive at the park entrance.

At the security and ticket check in the new Technoland, you are given Google-esque glasses. You and about 20 others are escorted to another room, where you are given a brief on how to work your new technology. After you've synced up your phone to your glasses and synced up to your friends and family, you head out into the park.

I've decided to be Princess Leia for the day, and my best friend is Luke Skywalker. Once we enter the park, we are no longer on planet Earth. We are on the planet Tatooine, stopping for fuel for the Millennium Falcon. My little sister is in the same physical park as me, but what she is seeing is completely different. My little sister just stepped into Wonderland. She is dressed in her blue dress, which she can see when she looks down at her shoes. She is Alice, and she is in the garden of the Red Queen. Over to her left, she can see the croquet field all set up to play, equipped with flamingos and hedgehogs. Before she has the chance to run off to the croquet field, we all decide to follow our itinerary and head to the *Star Wars*-themed roller coaster—also according to my app, there is no wait time! We scan our wristbands and breeze through the express line and immediately arrive at the loading zone of the ride. My sister and I hop into the front row, and my parents behind us. I see a notification on my glasses that says Syncing Star Wars Roller Coaster. I'm seeing the same thing as I did before because I'm already in Star Wars Land, but now my sister can see what I see. Once my glasses have loaded, the conductors come around to check our seat belts and we're off! Zooming through hyperspeed to try to outrun

Darth Vader during a meteor shower! As we pull back into the loading station and hop out of the seats, our glasses begin syncing to the outside of the park again, so my sister is back to seeing Wonderland.

After about an hour or so in the park, things start to pick up. More people come flooding through the gates, and it's almost time for a midmorning snack, but my sister wants to ride Alice through the Rabbit Hole, which features the most insane drop. I open my app, and it tells me that the wait time is 30 minutes! "No way!" I say. But fear not, my app has already suggested other activities for us to do until the wait time dies down. We can go play croquet or head to our favorite churro cart on our way to the Alice in Wonderland ride! We decide on the latter. After our experience buying a churro, our Google Glasses recognize our pleasant experience at this particular cart and remember to suggest it next time we're near. Then my sister spots a giant maze and pulls me toward it. I see it too! But to me, the walls of the maze aren't made of leaves, like they are for my sister, but instead are metal. I have to enter this maze to find critical information about the Death Star. My sister is trying to follow the White Rabbit, who keeps popping his head around corners, either helping us get through or leading us to dead ends. Once we reach the end, we both take off our glasses to see a beautiful hedged maze, large in scale, and flowers all around us. We put our glasses back on, and because we 3D scanned ourselves before we got to the park, we each take a picture in our lands, my sister as Alice and me as Leia. Our app adds it to a photo album for us to look at, add geofilters, and upload to social media later. I'm having the best day ever.

This above example details technology that does not yet exist but I believe will in the next five to ten years. Some parks have been implementing the customizable experience via the VIP package, but if you're a regular Joe like me, you've had to customize your experience the old-fashioned way, through multiple visits.

In Technoland, you can choose to be anyone you want to be, and this kind of experience is tailored to the millennial generation, who are constantly looking for more personalized experiences. They want something they can change and make theirs. Social media is all about the personal experience. Uploading images of ourselves makes us inherently self-centered because of, as one study suggests, "voluntary participation" in a "virtual reward structure."[3] Meaning young generations willingly subject themselves to public judgment, and they are looking further inward to get outside validation. Because this process is subtle but highly permeable in our daily lives, the younger generations expect the ability to continue "participating" in other aspects of their lives.

The personalized aspect of the experience narrated above will attract repeat customers because they can experience a new character each time they come. The sense of mystery in the multitude of other options available will leave the customer with the need to know who else they could be within the park walls, not to mention that a VR and AR experience is completely immersive. The success behind Harry Potter World, Pandora, and other immersive lands is already evident and will only continue to get better over time. Immersive lands will be the new norm for themed entertainment, and VR and AR will be a major help in the realm of cost effectiveness.

Okay, back to 2018. Is this technology feasible? I talked to expert in theme-park technology Jason Rauhoff about the possibility of the experience above, and he said that this could be feasible in the "next five years solid." He projects this technology to come quickly because the "backbone is already developed." By that he means that the baby steps of the virtual- and augmented-reality experiences are already made. For example, mobile augmented reality, which is the overlaying of digital information viewed through your camera phone—in other words, things like Pokémon Go—is already developed, and even on a more basic level with 3D and 4D rides.

My idea for the future just expands the idea to make a more immersive and less obstructed experience. So instead of having to hold your iPhone in front of you or wearing a heavy headset and portable computer, you get a smaller, lighter pair of glasses, or even no headset at all. This creates a more comfortable experience, larger in scale, and more immersive. Jason Rauhoff agreed that the early models of the experience that I hope to see in the next five years will probably be a Google Glass-type of experience since new models come out regularly.

A step toward this increasingly immersive future is the *Star Wars*-themed hotel experience Disney plans to open in 2019 in Orlando. If you haven't heard yet, Disney has basically made *Westworld* come to life, complete with droids and hotel "cast members," not workers! I geeked out when I heard about the new hotel idea. The Galaxy's Edge is an incarnation of my hopes and dreams for theme parks. When I was in The Wizarding World of Harry Potter, the whole time I was thinking, "I need more," and Disney brought the most. Galaxy's Edge, including its hotel, allows you to live your weekend as a full-blown character, complete with costume. You will leave planet Earth and stay upon your spaceship that sets sail to the Galaxy's Edge. "From the second you arrive, you will become a part of a *Star Wars* story," Bob Chapek, Disney's CEO, states."[4] The hotel will include "experiential multi-day adventures, views of space out of every window and full-on costume cosplay for its guests."[5]

There are still very few details released about the park, but all I need to know is that it's in the works. When you visit a place like The Wizarding World of Harry Potter, it's an incredible experience because you're living as the characters you love. However, the second you step outside of it to get back to your hotel room, the illusion ends, and you know it's not real. With the new *Star Wars* concept, your experience continues until you decide to head back to planet Earth.

One particular detail that has been released about Galaxy's Edge that I'm excited about is the Cantina. There's speculation about the creation of Mos Eisley's bar from Tatooine first featured in Episode IV. Her cantina is an iconic part of A New Hope, and Mos is one of my favorite characters! I remember watching the movies and thinking how amazing it would be to go there and meet her. I'm sure my dream will come true in 2019, but the characters in Star Wars land definitely won't be your average character meet and greets. They won't be giant felt versions of the ones you see on TV. They'll be much more lifelike, and you will actually be able to barter with them for some of the "local delicacies" at the street markets.

This hotel takes immersive lands to the next level. Once people experience the fullest level of immersion possible, there will be no going back. When Walt Disney imagined the future, he would have expected us to actually be able to take people to other planets by now, but this is about as close as we can get, and I think Walt would look at this project and smile.

Tomorrow is a heck of a thing to keep up with.

—WALT DISNEY

When Walt Disney opened Tomorrowland, his technology was already outdated. This might be the case forever with theme-park technology because it's always being improved, but Walt Disney would want us to strive for perfection, to look beyond the future in order to continue inspiring delight in other people.

CHAPTER 8

TOUR GUIDE

———

You're walking through Diagon Alley, and you come up to Weasley's Wizard Wheezes storefront. You plant yourself in front of the window that houses a toilet bowl. You wave your wand in an upside-down *D* motion while saying "descendo," and *FLUSH!* The flag sticking out of the toilet circles down the drain. You smile at your accomplishment and move on to the next interactive-wand destination.

"Alice, pull up my map of Diagon Alley." In your view appears the map of Diagon Alley. It looks similar to the Marauder's map from the movie.

"Alice, show me where the closest spell location is." A large red arrow appears, telling you to make a right after the Weasley's store into Carkitt Market. When you make the right, the fountain in front of you is highlighted like a halo surrounds it. You point your wand at the fountain and perform the correct motion while saying "Aguamenti," and voila! Water comes out of the fountain for you to drink from.

Once you've made your way through the Carkitt Market, you want to hit Escape from Gringotts. Your glasses are automatically scanned for your Express pass, and you walk right through. As you're waiting in line, you decide you'll want to grab some lunch afterward.

"Alice, please make a reservation at the Leaky Cauldron for lunch."

"Your reservation will be ready in 30 minutes. The line you are currently in is 20 minutes, so you may not have enough time to get to your reservation. Would you like to make a different reservation or keep the one you have now?"

"I'll keep my reservation. I think I can make it. Thank you."

After your ride, you head to the Leaky Cauldron just in time for your reservation. During your lunch of butter beer and fish and chips, you decide to head over to Hogsmeade once everyone has finished.

"Alice, how long is the Express line for the Hogwarts Express?"

"The Express pass line for the Hogwarts Express is currently 15 minutes. Would you like to reserve a place in line now?"

"Yes please."

"Okay. Your place is reserved. Please arrive at the train in 15 minutes."

"Thank you."

After everyone is finished with lunch, you make your way over to the train. You're given your cart, just like Harry had when he pushed through the portal. Through your glasses, you're seeing just a brick wall, and you're told to run straight through it, just like Harry and Ron did. But it's a bit terrifying staring at this brick wall and having to run full speed at it.

"Trust in your wizarding skills," says the train conductor.

In your head, you count, "one, two," rocking back and forth as you say each number, "three!" Off you go, charging straight at the wall. You close your eyes as you get close.

"Phew! I'm through," you think.

Once through the portal, you are ushered to your cabins on the train, where you're actually able to move about and buy some candy off the trolley!! The chocolate frogs are actually jumping away from you!

But, suddenly, the room gets cold, the windows start to get covered with ice. You breathe out and can see your breath.

"Dementors!" screams the terrified trolley attendant. She looks directly at you. "Can you perform the Patronus spell?"

"I, I think so." You nervously hope.

"Here they come," the attendant whispers in a terrified hush.

The window of your cabin flies open, and in pour two dementors coming straight at you. Out of instinct, you grab your wand, point it at them, think about your most happy memory, and yell, "EXPECTO PATRONUM!!!"

Immediately, light flows from your wand and expels the dementors from the train. The lights come back on, the temperature is back to normal, and Harry comes out from his cabin. "Good work, mate! You saved us all!"

Once off the train, you think about how awesome it will be to tell all your friends later about how you stopped the dementors.

"Alice, did you capture all that stuff from the train?"

"Yes, the whole train-ride video is saved in your camera roll."

Excellent. Now you can show your friends your awesome wizard skills.

"Alice, are there any secret spell locations in Hogsmeade?"

"There are two secret locations that are not on the provided map. One in the alleyway before the entry point of Hogwarts and another behind Honeydukes."

"Take me to the closest one."

The large red arrows appear in your vision. You follow each subsequent arrow to the alley behind Honeydukes.

"Okay, Alice, now what do I do?"

"Point your wand at the lollipops, wave your wand in this motion." Then the lowercase u shape appears in your vision, showing you how to perform the spell. "While saying 'wingardium leviosa.'"

As you trace the motion of the u and say the magic words, one of the lollipops begins to levitate, and you can even control its movements!

"Thank you, Alice. Now where can I find some butter beer?"

*　*

The above experience is half based on my actual experience at Harry Potter World and half what I wish had been included. When I actually rode the Hogwarts Express, I was stunned by the beauty and realness of what was happening just outside my window. But at the same time, I wish I could have moved about the trolley, and I fantasized that the same, sweet candy lady from the movies would walk by and say, "Anything from the trolley?" Then I could hop out of my seat, open the cabin door, and purchase a chocolate frog.

Reporter Amy Ziese describes the Hogwarts Express as "arguably the most immersive experience you can have in The Wizarding World of Harry Potter." Imagine if we could make it even more interactive!

The Google Glass technology could be a real game changer. Throughout the entire park, you wouldn't have to consistently take on and off headsets and goggles, breaking the illusion. Instead it would be a seamless experience. However, the technology needs some improving before it can have the kind of effect like on the Hogwarts Express.

BACKGROUND ON THE TECHNOLOGY

The first Google Glass was developed in 2013. By 2014, Google sent out the first glasses to 8,000 Google "Explorers" who would be the first to test out the new technology. However, upon initial review, it was described as, "useful, but overpriced and socially awkward."[1]

Glass works similarly to how your iPhone works. In fact, on the side of the glasses contains all the hardware, like Bluetooth, the processor, storage, and more. On the front of the glasses are the display screen, called a heads-up display, and a camera. The glass will "wake up" to the command, "okay, Glass." From there, you can do anything like search the web, take pictures, respond to texts and emails, etc., all through voice commands similar to Siri. But you can also toggle through the main screen using the right arm of the glasses like you're swiping on your smartphone.

One question I kept having during my research was whether there was a difference between the smart-glass technology and augmented reality. As I read more and asked more, I discovered that the glasses are a tool of augmented reality. Google Glass is similar to your smartphone, and since your smartphone can use AR apps, then the glasses have a similar capability. However, the question then becomes about the experience: Can the glasses produce the same realistic effect as the AR goggles? Where Google Glass was designed for the consumer's everyday life, the same smart-wear technology has applications in other industries. In fact, after the launch of Google Glass, a multitude of other companies saw the potential for wearable tech and started to build their own versions.

For example, Epson's Moverio BT-200 Smart Glasses use both lenses to project its display, which means twice the virtual screen size than that of Google's. "With a front-facing camera and motion tracker, the BT-200 is a premier development platform for apps of the future and hands-free scenarios. These binocular, transparent smart glasses open up a whole new world in entertainment, manufacturing, medical science, and more."[2]

PROMISING APPLICATIONS OF SMART-WEAR TECHNOLOGY:

- Health care
- Manufacturing
- Entertainment

INTRIGUING EXAMPLES OF SMART-WEAR TECHNOLOGY:

- **Smart glasses for athletes.** Recon Jet are smart glasses that display performance statistics while you're exercising. Without having to look down at the handlebars, bikers can see their speed, heart rate, and other important statistics. Also, you can receive texts and calls, so you can maintain contact while you're out on a run. You can also take pictures and consult the map in case you get lost.[3]

- **Smart glasses for manufacturing airplanes.** Making airplanes like Airbuses is very complicated, with confusing maps and detailed blueprints that are difficult to decipher. Accenture developed industrial-wear smart glasses in collaboration with Airbus to improve productivity and accuracy when manufacturing their airplanes. I know I would feel more comfortable knowing the manufacturers of my airplane used the latest and greatest technology to make the safest airplanes.[4]

- **Smart glasses for the visually impaired.** Aira is a smart-wear technology designed to help the blind navigate the world around them more effectively. Where the cane helps them to determine what is right in front and below their waist, it is still difficult sometimes not to bump into things above them, like tree branches. But Aira does not replace existing assistance systems, instead enhancing them by allowing the user to call an agent who can parse the environment for them and even describe the items on a menu at the restaurant.[5]

CURRENT APPLICATION IN THEME PARKS

No smart-glass technology is currently being used in theme parks; however, smart wear is slowly being incorporated into the parks. For example, Universal has started implementing virtual lines at their Volcano Bay water park. Customers are given Apple Watch-like wristbands called TapuTapu that act as their park tickets, credit cards, Express passes, and ride-reservation concierge. The wristbands allow customers to reserve their place in line ahead of time and then alert them when their turn is coming up.

When interviewing Justin Schwartz, the Director of Engineering and Safety at Universal Creative, I asked him what major complaints customers have. He told me that lines were the number-one complaint amongst customers. People pay good money to come to the parks, and they feel it's a waste of their time and money to be standing in lines for an hour or so. Even Jason Surrell, one of the Universal creative directors, said, "We've known for years that waiting in line is one of the biggest dis-satisfiers in our guests' day."

One thing that always bothered me about the Fastpasses at Disneyland was that not all rides had them, and also that I still had to wait in line! Everyone who got the same time on Fastpass was also showing up, so we were still waiting for 20 minutes to get on the ride. The Express passes at Universal were the best purchase I made, besides my Slytherin sweater, because I was in line for a maximum of 10 minutes for every ride. Universal's TapuTapu just took things to the next level.

Interestingly, they got their inspiration from airports, where you reserve your seat in advance and then show up for your flight when it's time to board.[6] According to customer reviews, however, TapuTapu did not work as well as Universal predicted. First, most people did not know how to use the technology, let alone that the only option was to reserve ahead for their spot in line. And even if they did reserve, they still had to wait in lines, sometimes up to half an hour! So unfortunately, the lines are still an issue, but it's certainly a step in the right direction.[7]

Even though smart-wear technology has not yet gotten to the theme-park industry, I believe it could have enormous application. In the Harry Potter story described above, each guest has their own personal tour guide to help them navigate the park and act as their concierge, and it's even equipped with fun facts and secrets about the park. Furthermore, using the glasses as virtual- and augmented-reality headsets, as described in previous chapters, could mitigate costs for robots, infrastructure, and ride repurposing. As a frequent flyer in the theme-park world, I know smart-wear technology would greatly enhance my experience.

CHAPTER 9

SELF-DRIVING CARS AND ROBOTICS

———

"Backpack, check. Wallet, check. Sunscreen, check. Tickets, check."

"Bow and arrow, check." I smiled as I interrupted my mom's mental checklist.

"All right then," she shook her head, "off we go."

As we headed to our local pickup station, I looked through the window at our earthly planet one last time before I headed off to Pandora.

"Why is peace among our world so much more difficult than in Pandora," I wondered.

We pulled into the pickup station for Disney guests, and as we parked our car, my palms started to sweat from excitement. I wondered what it would be like to be a Na'vi for the day. What kind of creatures would I encounter?

The excitement was building up even more as our bus pulled in. My mom and I, and about 20 other eager families, piled into our shuttle that would take us to Disneyland for the day. As we walked to our seats on the bus, no one said hi to the bus driver, since bus drivers don't exist anymore.

"Back in my day, we would always say 'hello' and 'thank you' to our bus drivers." I rolled my eyes as my mom began to reminisce on the good old days.

The Disneyland bus station uses self-driving buses to drive LA locals to Disneyland, which is great because we save money on gas and it's so much better than driving ourselves to Anaheim.

I watched the trees blur by as my mom read her new book on her iPad. I was counting down the seconds until I could land on Pandora, my favorite planet, and live my favorite movie, *Avatar*.

After about an hour, we finally arrived. We stepped off the bus and were shepherded toward the ticket and security check. Our bags weren't checked by regular people; instead, I was being inspected by real Na'vi that talk, and think, and move.

"Pick your jaw up off the floor, honey. They cleared us to go in."

I followed my mother's instructions and headed into the park. I saw Jake Sully and Neytiri! This was so cool.

"Back in my day, our favorite characters were in big felt suits and couldn't even talk to us!"

"That's so lame, Mom. Why even bother with the characters if you can't talk to them?"

I ran up to Jake.

"Jake! Oh my gosh, you're really here. I'm such a huge fan!"

"Hi, it's nice to meet you…"

"Oh, Maddie, my name's Maddie."

"Maddie, pleasure."

"How did you know it was the right thing to disable that bull-dozer? Weren't you scared?"

"A little, but sometimes the right thing to do isn't always the easiest, but if you follow your instincts," he pointed to my heart, "then you'll always know the right choice."

"Wow, thank you so much, Mr. Sully."

"Call me Jake."

Life. Made. I was on cloud nine for the rest of the day.

BACKGROUND ON THE TECHNOLOGY

This is what theme parks could look like with the application of self-driving cars and other artificial-intelligence robotics. These two technologies are somewhat more in the future, while virtual and augmented reality are happening now. Nonetheless, I think these two things could have interesting applications to the theme-park industry. But first, a little background on the technologies.

Self-driving cars, also known as autonomous vehicles, began in 1925 with Houdina's radio-controlled car that was demonstrated during the New York World's Fair. The "American Wonder" was operated by an antenna that would get information from a second car that controlled it. By driving behind the Wonder, the second car would send out radio impulses that would communicate with the circuit breakers in the car, which operated the movements of the car.[1]

In 1960, Ohio State University's Communication and Control Systems Laboratory set a project in motion that would develop driverless cars that would communicate with sensors in the streets it drives on. This technology would be considered by the Transportation and Road Research Laboratory in the United Kingdom, and used in the Netherlands as people movers for Schiphol Airport in 1997. [2]

Today there are a multitude of brands that boast a driverless car, such as Tesla, Audi, Lincoln, Cadillac, etc. Most of these cars function by using cameras, radars, and sensors to create a 360-degree view of the car. While there is still some controversy surrounding the safety of these vehicles, I believe that the driverless car is an inevitability.

INTRIGUING EXAMPLES OF DRIVERLESS CARS:

- **Uber's self-driving cars.** Uber has recently struck a deal with Volvo in efforts to create a self-driving rideshare fleet. In fact, they deployed a fleet in Arizona as one of the first testing grounds (San Francisco being the first, but not successful) for their fleet. There will still be two safety drivers in the front in case of a need for human assistance. But eventually, there would be no more Uber drivers, just the Uber. A friend of mine who lives in Pittsburg had an experience with the self-driving fleet there, and she recalls a large screen so you can see the navigation and a safety passenger for the "just in case"![3]

- **Parking.** No longer will we have to spend what feels like hours searching for parking spots; instead, our autonomous vehicles will drop us at our destination and head off to find a spot to park and wait for us to call it with our iPhones when we are ready to depart. No more parking tickets, tows, or people who park in two spots, leaving you without one.

- **Kids could drive themselves to Karate practice.** Since, in the likely future, you won't need a license to drive a car that drives itself, kids will be able to drop themselves off and pick themselves back up again from their extracurricular activities, even to and from school! [4]

CURRENT APPLICATION IN THEME PARKS

Walt Disney World is planning to test out driverless cars in Florida. They will first test their fleet on their employees, as they already use a massive fleet of regular buses that transport their employees. If those tests succeed, they will allow guests to use the buses as well. The *LA Times* projects Disney's deployment as early as this year. Despite California's growing number of driverless cars, such as Tesla, Florida has already set in motion some of the most welcoming laws for the autonomous vehicles.[5] The U.S. State Department has named Florida as one of the ten autonomous vehicles proving grounds, and Epcot will begin with a demonstration of the cars in the next year or so.[6]

The use of driverless cars in theme parks could be a great help in making it easier for customers and employees to get to and from the park. Currently in Los Angeles, the Surfliner will take citizens to Anaheim, with shuttles to Disneyland. But if Disney had their own fleet of driverless buses, it would eliminate the need to change buses, making the process more streamlined and efficient. Disney would also save money since they wouldn't have to pay wages for the drivers.

In addition to the futuristic self-driving cars, artificial intelligence in the form of robotics could also have a notable impact on the theme-park industry.

Remember earlier, when Maddie was able to talk to Jake Sully in Pandora? Thanks to the help of artificial-intelligence robots, characters will no longer be in felt suits.

BACKGROUND ON THE TECHNOLOGY

The history of artificial intelligence has its roots in ancient history, starting when Aristotle invented syllogistic logic, which is the first formal deductive-reasoning system. (Reasoning systems within the computer play an important role in the authenticity of the artificial intelligence.) Fast forward to the eighteenth century, when artificial intelligence was no more than a toy. Few scientists saw the application of AI past toys like the Turk, a mechanical chess player. However, by 1956, as curiosity increased, there was finally enough interest in the possibilities of thinking machines to declare AI an academic discipline. But before that, the initial research into artificial intelligence stemmed from brain research that came out in the thirties, forties, and fifties. Scientists discovered that the brain is a system of electrical impulses, which inspired Alan Turing to publish his paper on the thinking machine. Turing thought, if the brain was a series of electrical impulses, then the computer was not so different, and everything could become ones and zeros. Turing believed that if a machine could carry on a conversation with a person without that person knowing they were talking to a machine, then that machine could "think." Alan Turing's Turing Test was really the beginning of artificial intelligence. After Alan Turing's paper, a group of scientists held the Dartmouth Conference of 1956. This conference gathered many scientists, including two from IBM, to discuss this assertion: "Every aspect of learning or any other feature of intelligence can be so precisely described that a machine can be made to simulate it."

In the years after the conference, computers were beginning to solve algebra word problems and even learning to speak Spanish! However, in the 1980s, a form of AI called "expert systems" emerged, which are programs that solve problems about a specific area of knowledge, using logical rules that are set by experts. For example, Edgar Feigenbaum and his students developed Dendral, which was able to identify compounds from spectrometer readings, and in 1972, MYCIN was able to diagnose infectious blood diseases. All of these expert systems relied on the expert knowledge of their particular field.

One of the biggest milestones in the history of AI came in 1997, when IBM developed the first computer able to beat the reigning chess champion, Garry Kasparov. Deep Blue was reportedly capable of processing 200 million moves per second.[7]

In the twenty-first century, access to larger amounts of data and faster processing speeds of computers has allowed for rapid development of AI. We are all familiar with Siri, whom Apple developed in 2011. Then with movies like *Her* and *Transcendence*, AI has become a phenomenon. In the movie *Her*, Theodore falls in love with his new operating system. She is intuitive and thoughtful with him, and things between them even get intimate, but he is heartbroken to find out that she is also talking to thousands of other users with the same OS. Just a couple years after *Her* premiered,

Dr. David Hanson of Hong Kong's Hanson Robotics showed off Sophia. The high-functioning, beautiful robot, Sophia, complete with skin graft, strikes a remarkable resemblance to Audrey Hepburn. But Sophia isn't just a pretty face; she is also witty and adaptable and intuitive. CNBC's Andrew Sorkin interviewed her and reported that she "displayed the ability to understand the human emotion of 'uneasiness' and 'doubt.'" Science-fiction movies are slowly becoming our reality, especially in the world of AI. In fact, in 2014, there was a call for a new type of Turing test to be able to keep up with the fast-changing pace of modern AI. [8]

PROMISING APPLICATIONS OF ARTIFICIAL INTELLIGENCE:

- **Health care.** Johnson & Johnson's Sedasys system has FDA approval to administer anesthesia to patients in routine operations. This machine replaces human anesthesiologists, saving costs, as now only one doctor oversees multiple machines.

- **Manufacturing.** Manufacturing companies have been on the AI bandwagon for a while, using robots to assemble and package products. In the future, these robots will be able to assemble more complicated items, like cars and even houses!

- **Transportation.** As discussed earlier, self-driving cars will change the public-transportation industry. But in order for these cars to work, they need to be able to process and react to different and complicated material quickly, which is where AI comes in. Driverless cars are predicted to eliminate accidents, tickets, and even traffic congestion.[9]

INTRIGUING EXAMPLES OF ARTIFICIAL INTELLIGENCE:

- **Music humans have never heard before.** NSynth, not to be confused with NSYNC, is an application developed by Google that has the capability to produce musical sounds that humans have never heard before! NSynth uses a large database of sounds and runs them through a neural network, which then analyzes the notes that stand out. Then the neural networks create new sounds by melding existing sounds.

- **The Wi-Fi that just gets you.** MIT is currently developing an application called EQ-Radio that in connection with your Wi-Fi network can detect your emotions. Scientists can detect your heartbeat by bouncing signals off the user and then running those signals through an algorithm that could detect someone having a heart attack or monitor how calmly your newborn is sleeping.

- **"The Machine That Dreams."** Google's artificial neural networks process the immense number of images on Google's servers. But at night, their server dreams, just like we do. They first begin by teaching the computer what things look like. For example, it teaches the computer what a spoon looks like by showing it hundreds of images of spoons, and the computer analyzes the major features that make a spoon a spoon. That same process the computer uses to understand images can be reversed to generate images. However, instead of being able to produce a clear image of just one spoon, it would create these beautiful, psychedelic, Dali-esque images. Interestingly, this could be an argument against Elon Musk's worst nightmare, because even though the computers are intelligent, they do not yet possess the capability to consolidate like humans do in the creative sphere. [10]

CURRENT USE IN THEME PARKS

Eighty-five percent of theme-park visitors from the US, UK, China, Japan, and Malaysia want artificial intelligence in their theme parks.[11] Facial recognition, fingerprinting technology, and palm-reading technology for ID verification is favored to make visits as trouble-free as possible. Universal Studios currently uses facial recognition for Express passes and fingerprint verification for park tickets.

Disney is also looking to use AI robots to replace the characters that we all know and love.[12] So, like in the example given early in this chapter, our children will be able to actually meet Jake Sully, not just wave at the guy in the suit.

I believe that AI also has the potential to replace the conductors of the ride. They would be able to assess a problem with a roller coaster and be able to fix it on-site, instead of having to wait for the mechanic to come and look at the ride and shut it down temporarily. Instead, the robot would be able to find and solve the problem more quickly and efficiently.

However, with both artificial intelligence and self-driving cars comes a downside. The potential for job loss in any industry is imminent. In 2013, Oxford University conducted a study that showed in the next two decades, 47 percent of US jobs would be replaced by robots.[13] Engineers, scientists, and economists are becoming the most sought-after employees, according to Bill Gates, because their skills are the most valuable in a world full of robots. Nonetheless, it's becoming a fact of life that robots will be taking away some of our jobs, and the theme-park industry is no exception.

CHAPTER 10

CONCLUSION: IT'S (STILL) THE LITTLE THINGS

—

Enjoy the little things, for one day you may look back and realize they were the big thing.

—ROBERT BRAULT

As we sat on the couch, watching the trolley scene from *Meet Me in St. Louis*, I looked over to my mom and smiled. A full three months of time together, the most time we'd spent together consecutively since I'd first left for college, and as we both teared up at the thought of Esther not getting to be with John because of their move to New York, I was confident that this was a little moment together that would be one of those "big things" in my life that I would remember.

Tomorrow she'd take me to the airport, after perhaps one of the last times we would spend time of this length together, given the fact that normally, she lives in China.

*   *

Walt Disney gave himself less than a year to build Disneyland. But throughout the experience of building his dream, he never sacrificed the details.

"When we consider a new project, we really study it—not just the surface idea, but everything about it."—Walt Disney

Disney was not interested in just creating a surface idea; instead, he wanted to build the whole idea. Meaning it wasn't enough just to have roller coasters and shows—he had to have every aspect of the stories the rides told. He wasn't satisfied with just teacups. The small details of each individual land were what transformed the 160 acres into a magical space that brought 50 million people through its gates in its first decade.

Why was the experience so radically different than any amusement park ever in history?

Tomorrow can be a wonderful age. Our scientists today are opening the doors of the Space Age to achievements that will benefit our children and generations to come... The Tomorrowland attractions have been designed to give you an opportunity to participate in adventures that are a living blueprint of our future.

—WALT DISNEY

Autopia was the main attraction of Tomorrowland on opening day in 1955. Guests could drive cars as a preview of what highways would look like tomorrow. You could actually walk into your futuristic home and open the plastic cabinets and plastic drawers in the kitchen to get out your plastic plates, as a peek into the amazing applications of the technology that people at the time did not realize. You could also ride in a space shuttle and feel what it's like to walk on the moon.[1]

In Tomorrowland, guests were confronted with the products that Walt felt would change the future in such a seamless way. The accents on the buildings were all painted silver. Everything had a rocket theme, and even the futuristic Monsanto House had a modern design. All these details led the customers to forget all about the present that existed outside the magic walls. People left Disneyland feeling like they participated in a time-travel experiment because, for the day, they lived in their future.

Walt's goal for Adventureland was "to create a land that would make this dream reality, we pictured ourselves far from civilization, in the remote jungles of Asia and Africa." Frontierland brought us to not a different place but a different time.

All of us have a cause to be proud of our country's history, shaped by the pioneering spirit of our forefathers... Our adventures are designed to give you the feeling of having lived, even for a short while, during our country's pioneer days.

—WALT DISNEY

Both Frontierland and Adventureland exposed customers to a way of life that was unlike that which they lived every day. That point could be applied to all of Disneyland, but particularly in these two lands because it incorporated a sense of culture. The moment you walked into Adventureland, you were confronted with tribal masks signaling your departure from the last land you came from. The thatched roofs, the wooden poles with skulls on them, and even the Adventureland banner made from sticks and rope all gave a sense of the impending adventure, in which you were transported to a safari in Africa, confronted by hippos and elephants. Frontierland also brought that sense of adventure, but instead of being set in Asia or Africa, Disney brought back the American frontier. A ride on the Tom Sawyer steamboat transported guests back into the book and movie. The petting zoo designed as a backyard pen, the Grand Canyon-esque design of Splash Mountain, all help to create the effect of the American frontier.

What youngster... has not dreamed of flying with Peter Pan over moonlit London, or tumbling into Alice's nonsensical Wonderland? In Fantasyland, these classic stories of everyone's youth have become realities for youngsters—of all ages—to participate in.

—WALT DISNEY

Fantasyland was created with the intention that when you wish upon a star, all your dreams will come true. The white, light-blue, and pink color scheme coupled with the old European brick building design helps to set the stage for the Disney princess films. The happy, light orchestra that plays in the background gets you into Snow White's mindset that life is beautiful and happy. Families could come tour Cinderella's castle, engage with their favorite characters on Mr. Toad's Wild Ride, and even spin around in giant teacups that could only exist in the world of the Mad Hatter.

I asked Mohamed Newera, an Imagineer for over 20 years, if he had ever been to Alton Towers. I explained how there was no overarching theme to the lands or to the park in general. Mo told me that the park itself was very old, but they saved a ton of money because there wasn't the type of detail I was used to at Disneyland or Universal Studios. This discussion sparked the memory of having to deal with hiding a main water pipe in Disneyland Paris. He and his team had to take a rainwater pipe and hide it inside a steel column. He recalled how hard it was to take the steel column, lift it up, slip through it the rainwater pipe, work the rainwater pipe down into the drain, and then place the column down and connect the rainwater pipe's head to the bottom of the column. "It's a killer," he said. "Disney spent so much money to hide that pipe. Today it can be decorated or painted and might not be such a big deal. Spend the money where the guests can see!" For someone working in the industry, he recognizes that a lot of the money is spent in the details because, at the end of the day, it's the most important.

When I saw *Fantastic Beasts and Where to Find Them* for the first time in the theater, I remember during the banquet scene, I looked to see if there were champagne glasses on the tables. I wondered if someone maybe had slipped up in the Prohibition-era movie and accidentally included alcohol at the event. They didn't, fortunately for me, as that seemingly tiny little mistake would have ruined the movie for me. It would be like if you went to Colonial Williamsburg and the lady churning butter said, "What's up?" to you!

More than these discrepancies simply being confusing as a reader and a viewer, it would've disrupted the magic. That's why Walt made sure even all the trashcans on Main Street looked era-appropriate. This kind of magic is what draws people to come back time and time again. It's why we're seeing the increase of fully immersive lands, because as customers, we want more! We want more magic, more moments of escape! The idea is to make the impossible possible by transporting us to places we could never go.

And at the very front of the park:

for those of us who remember the carefree time it recreates, Main Street will bring back happy memories. For younger visitors, it is an adventure in turning back the calendar to the days of grandfather's youth.

—WALT DISNEY

The old movie theater, the horse-drawn trolley, the early twentieth-century small-town American architecture, and the smell of the confectionary and old-school hot dogs all aid in turning back time. Main Street was particularly important to Walt because it was a recreation of his childhood home in Marceline, Missouri. The very first thing the guests encounter is a return to Walt's childhood, and he would argue, a return to childhood in general. Walt was a humble, ordinary guy whose perseverance to live out his dreams brought him to the big time. But he never lost sight of the "why" behind his projects. Walter Elias Disney is such an inspiration to me. He took ideas from his past growing up in Marceline, and despite the abuse he suffered by his father, he chose to see the good in life instead of dwelling on the bad.

He made Main Street in Disneyland as his way of recreating his past, erasing all the negative and leaving only the positive. He did so in such great detail that it's no wonder he liked to live in Disneyland over the weekend. His ever-present optimism pushed him forward through even the hard times of his career. He never let financial trouble, or company strikes, or war stop his vision for the world. He was constantly told throughout his life that he would never make any money drawing cartoons. His father told him to get a real job at the jelly factory because dreams are just dreams and they will never come true.[2] He believed in himself though, and he knew that by embracing the nostalgia of the past and letting go of the painful memories associated with the past, he could create his idea of a better tomorrow. We can all, in some way, relate to Walt. We've all been told we can't do something, or that our dreams are unrealistic. When I was applying for college, I was told by my college counselor and one of my teachers I was very close with that I could never get into Wellesley College, and even if I did, I would never survive. To them, I wasn't smart enough, or driven enough, or dedicated enough, but like Walt, I proved them wrong, and I "had a wonderful time doing it!"[3] As much as he made Disneyland for youngsters, he also made it for the child within us all. Immediately the guests are asked to return to a time when life was without stress and worry, to when life was full of wonder and possibility.

While Disneyland offers roller coasters and Ferris wheels, Walt's theme parks are not just about the thrill; they are just as much about the nostalgia.

Stepping inside the Disney and Universal movies and other books when I visit the parks takes me back to the moment when I was reading and watching them for the very first time. Walking through Weasley's Wizard Wheezes in Diagon Alley at Universal Studios, I remember that looking up toward the ceiling and seeing the little robot biking across the tightrope made me recall the exact scene from *Harry Potter and the Half-Blood Prince*. It was like I was in the movie. I looked up toward the ceiling, expecting to see Ron Weasley looking down at me. Every single detail was present, from the little robot riding its bike above me to the endless shelves of candy that just kept going up. When I read my books at night, setting aside my own world in a box on my dresser table, and delving into somebody else's world, escaping from my reality is what theme parks bring to life.

For me, watching TV and reading books is like putting on this really cool suit that allows me to travel between realms and worlds. I step into something that I'm not familiar with, and I can experience something and be someone completely different. Even as a kid, I tried so hard to be Matilda. I would sit and stare at a pencil for hours, trying to get it to move. I would focus really, really hard on the pencil, yelling in my head, "Move!" It never did, but I just attributed it to my lack of focus, never to the fact that it was impossible.

I heavily invest in my characters, so if the story doesn't end the way I had imagined, I get almost offended, because I felt a part of that world. All the details in books and movies are carefully planned out so as not to break the illusion. Theme parks, and more so immersive lands, need to be the same way. If all of a sudden, Harry Potter did something selfish, we would all be very confused because that is out of character. J.K. Rowling—and every other author—ensures that the plot, characters, themes, and motifs all fit seamlessly together in order to create a believable and relatable world while still maintaining a sense of awe and wonder.

PROTECTING THE PARK

Theme parks are different than amusement parks in one major way: Their themes are and must be protected from copycats.

Experts share that the most successful franchises are successful because of the intellectual property behind them.

Every industry expert I spoke with said the same thing: The most important thing is to start with a good story.

Walt Disney famously said, "I only hope that we never lose sight of one thing—that it was all started by a mouse."

Having gone to an all-women's college has taught me to think critically about Walt's films and how they affect young girls. But for me, I aspired to be Ariel, not because she married a prince but because she did the unconventional. She always felt like an outsider because she wanted to see the world, but she chased her dreams, and her strength allowed her to pursue and eventually live out those dreams. Cinderella inspired me to be kind, even to those who are not kind to me. Belle taught me to look beyond the exterior because everyone has their own stories, and it's important to think before judging someone. All of these movies were lessons Walt was trying to engrain in us as children, to grow up and make the world a better place. All of these feelings and emotions are what keep us coming back to Disneyland, and other theme parks. All inside the magical walls of the park, our childhood memories and dreams are vividly reproduced for us to enjoy.

Technology can certainly help cut some of the immense costs of all the details that go into making things like Main Street, from the store windows to the pavement on the floor. But it's important to remember that these experiences are nothing without the motivation behind them!

One of the best experiences I've ever had at an amusement park was at Alton Towers. I talked about it earlier as not having the kind of details that Disneyland or Universal has, but nonetheless, I had the best day there. All of their rides were surprising and elating. But my favorite part of the day wasn't even on a roller coaster.

On our way to ride another roller coaster, we happened to pass what looked like the sweeper from the TV show *Wipeout*. If you've never seen *Wipeout*, they always feature this contraption that has arms extending from a larger pole in the center. The arms spin around the central pole, and you have to jump over or under each of the arms. Alton Towers had a smaller, softer version of the one on *Wipeout*, but being such a huge fan of the show, I grabbed my boyfriend and told him we were doing this. After you take your shoes off, you each hop onto your individual foam podium that you use as your platform to jump over and duck under the sweeping arms. Also, it's worth mentioning that at this point in my study abroad, I had done zero exercise and eaten more pasta than you can imagine, so I was a little out of shape. Nonetheless, at the beginning, the arms were moving pretty slowly, and jumping over them was not really an issue yet. The hardest part was to accurately land back onto the podium after jumping over the arms. Then the conductor kicked it up a notch. I could not breathe, between the laughing and the physical exertion of jumping over these arms. There are few things funnier in this world than not seeing the arm coming before it just completely knocks you over. At this point, we had developed quite a crowd watching us humiliate ourselves. I had missed one of the arms and was preparing for the next one to come around, but I didn't turn around in time and got knocked down all over again. The crowd of people and my boyfriend were all laughing at my complete lack of coordination. By the end of our turn, I was completely breathless, lying on the floor of the sweeper. But for the rest of the day, I would chuckle at the thought of that experience.

The sweeper was one of my favorite moments at this theme park, and it wasn't even a real ride, not to mention it wasn't fairly technologically advance. The reason I had so much fun was because it was an experience that I had with my boyfriend and a bunch of complete strangers that made us all laugh so hard we cried. Those are the moments we remember. As much as theme parks are about the rides and the technology, they are also about the moments we share with our family and friends who accompany us.

For Disney, this is what it was all about. He wanted a place to spend quality time with his two daughters and, in general, bring joy to other families all over the world. Disneyland's

California Adventure Park features a *Finding Nemo*-themed attraction where kids can actually have conversations with Crush the turtle. The kids are placed in front of a screen that looks more like the glass window at an aquarium than a TV. Crush "appears to interact with the audience, answering its questions, making jokes, and otherwise acting like a real live character. In effect, it is live-action animation." One mother wrote a letter to Disney, thanking them for the experience they gave her son. Before meeting Crush, this little boy hadn't spoken in six years. When he met Crush for the first time in real life, he had his first conversation for a whole 10 minutes.

This is why I'm passionate about theme parks. It brings people's dreams to life. If immersive technologies can bring about more interactions like the one above, then theme parks should certainly invest in them. However, the most important thing to remember about the future of theme parks is that even though the technology will always be changing, the one thing that should remain constant is that it all starts with a good story.

ACKNOWLEDGMENTS

I could fill an entire book with just those whom I would like to thank for making this book possible. But I'll start with Eric Koester. Thank you for being available 24/7 throughout this process. Without your unwavering confidence and support I literally wouldn't have a book. To the entire Creator Institute team, thank you for spending the time brainstorming and revising. This process brought out the person I've always wanted to become. To my Mom and Dad, thank you for being patient. My path has not been straightforward, but a winding amalgamation of hills and valleys. Thank you for bearing with me while I took my own journey to discover where my path leads. To my big brother, Jordan, your exuberant confidence inspires me every day to be the best possible version of myself. Never stop being curious. And finally, to Chandler, my best friend and my love. Through this whole process, on my highest days you championed my progress and on my lowest days you were there, hand outstretched to raise me up. I am confident that I would not be here without you.

NOTES

___

PART 1

INTRODUCTION

[1] Ina Fried. "People Willing to Go Without Sex, Shoes and Caffeine Rather Than Give Up Their Cellphones." *All Things D*. August, 2011. http://allthingsd.com/20110808/people-willing-to-go-without-sex-shoes-and-caffeine-rather-than-give-up-their-cell-phone/(Accessed February 2018).

[2] Simon Maybin. "Busting The Attention Span Myth." *BBC News*. March, 2017. http://www.bbc.com/news/health-38896790 (Accessed February 2018).

[3] TEA/AECOM 2016 Theme Index and Museum Index: The Global Attractions Attendance Report. (http://www.teaconnect.org/images/files/TEA_235_103719_170601.pdf)

[4] Jean M Twenge. "Have Smartphones Destroyed A Generation?"

The Atlantic. August 2017. https://www.theatlantic.com/amp/
article/534198/ (Accessed February 2018).

CHAPTER 1

1 *American Experience*. "Walt Disney." Sarah Holt, Producer and
Director.

2 *American Experience*. "Walt Disney." Sarah Holt, Producer
and Director.

3 *American Experience*. "Walt Disney." Sarah Holt, Producer
and Director.

4 Gilder Lehrman Institute of American History. Carol Berkin,
Editor.
https://www.gilderlehrman.org/history-by-era/
development-west/timeline-terms/frederick-jackson-turn-
ers-frontier-thesis-0(Accessed February 2018).

5 Rainbow Mountain Stagecoach Ride. Disneyland 1956. http://
disneyexaminer.com/wp-content/uploads/2015/09/disney-
land-original-frontierland-stagecoach.jpg

6 *American Experience*. "Walt Disney." Sarah Holt, Producer
and Director.

CHAPTER 2

1 "Bakken: The World's Oldest Amusement Park." Entertainment
Designer. July, 2011. http://entertainmentdesigner.com/
history-of-theme-parks/bakken-the-worlds-oldest-amus-
ment-park/ (Accessed February 2018).

2 "Vauxhall Gardens 1661–1859." David Coke. http://www.

vauxhallgardens.com/vauxhall_gardens_briefhistory_page.
html (Accessed February 2018).

[3] "Mini Exhibit: Cairo in Chicago." The University of Chicago.
The Oriental Institute. September 2015. https://oi.uchicago.
edu/museum-exhibits/special-exhibits/cairo-chicago(Accessed February 2018).

[4] "New York Amusement Parks." PdxHistory.com. http://
www.pdxhistory.com/html/coney_island.html (Accessed
February 2018.)

[5] "The History of Knott's Berry Farm: From Roadside Pie Stand
to Theme Park." Entertainment Designer. September, 2011.
http://entertainmentdesigner.com/history-of-theme-parks/
the-history-of-knotts-berry-farm-from-roadside-pie-stand-
to-theme-park/(Accessed February 2018).

[6] *American Experience.* "Walt Disney." Sarah Holt, Producer
and Director.

[7] *American Experience.* "Walt Disney." Sarah Holt, Producer
and Director.

CHAPTER 3

[1] "Disney Tried to Invent a Timeless Tomorrowland. Here's
Where it All Went Wrong." *Theme Park Tourist.* Brian
Krosnick. January, 2016. https://www.themeparktourist.com/
features/20160103/31169/possibilityland-tomorrowland-dis-
ney-almost-built-and-failure-it-got-instead(Accessed
February 2018.)

[2] Pat Williams and James Denney. *How to be like Walt:*

capturing the Disney magic every day of your life. Health Communications, 2004.

3 Tomorrowland, 1955. https://www.themeparktourist.com/sites/ default/files/u235/peoplemover/Tomorrowland%201969.jpg

4 Monsanto House of Tomorrow, 1955. http://disney.wikia.com/ wiki/Home_of_the_Future

5 *American Experience.* "Walt Disney." Sarah Holt, Producer and Director.

6 "Customer delight from theme park experiences: The Antecedents of Delight based on Cognitive Appraisal Theory." *Annals of Tourism Research,* Pergamon, 9 Apr. 2013, www.sciencedirect.com/science/article/pii/S0160738313000406.

7 "Harry Potter boosts Universal Studios attendance; Disneyland visits slip." *Los Angeles Times.* Hugo Martin. June, 2017. http://www.latimes.com/business/la-fi-universal-atten-dance-20170601-story.html (Accessed February 2018).

8 "Why comparing Avatar with Harry Potter is a bad idea." *Theme Park Insider.* Robert Niles. March, 2017. http://www.themeparkinsider.com/flume/201703/5469/(Accessed February 2018).

PART 2
CHAPTER 5

1 "Disney's Haunted Mansion Is A Terrific Mess." Kevin Wong. *Kotaku.* April, 2016. https://kotaku.com/disneys-haunt-ed-mansion-is-a-terrific-mess-1773263620 (Accessed February 2018).

[2] "The Dark, Troubled History of Disney's Haunted Mansion." David Mumpower. *Theme Park Tourist.* August, 2015. https://www.themeparktourist.com/features/20150808/30470/ghoulish-history-disneylands-turbulent-haunted-mansion (Accessed February 2018).

[3] "History of Virtual Reality." *Virtual Reality Society.* https://www.vrs.org.uk/virtual-reality/history.html (Accessed February 2018).

[4] Morton Heilig's Sensorama. https://www.engadget.com/2014/02/16/morton-heiligs-sensorama-simulator/

[5] "History of Virtual Reality." The Franklin Institute. https://www.fi.edu/virtual-reality/history-of-virtual-reality (Accessed February 2018.)

[6] "5 Reasons Why 2017 is the Year Virtual Reality Becomes a Thing." Molly St. Louis. *Inc.* April, 2017. https://www.inc.com/molly-reynolds/5-reasons-why-2017-is-the-year-virtual-reality-becomes-a-thing.html (Accessed February 2018).

[7] "8 Crazy Examples Of What's Possible In Virtual Reality." Jonathan Vanian. *Fortune.* March, 2016. http://fortune.com/2016/03/22/sony-playstation-vr-virtual-reality/ (Accessed February 2018).

[8] "8 Crazy Examples Of What's Possible In Virtual Reality." Jonathan Vanian. *Fortune.* March, 2016. http://fortune.com/2016/03/22/sony-playstation-vr-virtual-reality/ (Accessed February 2018).

[9] "Wevr's TheBlu delivers a mesmerizing deep sea experience in VR." Dean Takahashi. *Venture Beat.* March, 2016.

https://venturebeat.com/2016/03/28/wevrs-theblu-delivers-a-mesmerizing-deep-sea-experience-in-vr/ (Accessed February 2018).

[10] Osso VR. http://ossovr.com (Accessed February 2018).

[11] Alton Tower's Galactica Rollercoaster. https://www.birminghammail.co.uk/whats-on/whats-on-news/alton-towers-new-galactica-ride-11053065

[12] "Inside the First VR Theme Park." Rachel Meltz. *MIT Technology Review.* Decemeber, 2015. https://www.technologyreview.com/s/544096/inside-the-first-vr-theme-park/ (Accessed February 2018).

[13] "Coming Soon to AMC Theaters: Virtual Reality Experiences." Brooks Barnes. *The New York Times.* September 2017. https://www.nytimes.com/2017/09/26/business/media/amc-theaters-virtual-reality.html (Acccessed February 2018).

[14] "AMC Strikes $20 Million Pact With Dreamscape Immersive for 6 VR Centers, Content Production." Janko Roettgers. *Variety.* September, 2017. http://variety.com/2017/digital/news/amc-dreamscape-vr-center-partnership-1202570835/ (Accessed February 2018).

[15] "8 Facts About The Pirates of the Caribbean Ride That'll Have You Saying, 'Yo-GO!'" Kelly B. *Disney Fanatic.* December, 2014. https://www.disneyfanatic.com/8-facts-pirates-caribbean-ride-thatll-saying-yo-go/ (Accessed February 2018).

[16] "13 Facts About Disney's Haunted Mansion." Alan Finn. *Mental Floss.* August, 2014. http://mentalfloss.com/

article/58272/13-facts-about-disneys-haunted-mansion (Accessed February 2018).

CHAPTER 6

[1] "Game Over For Virtual Reality?" *The Economist.* December, 2017. https://www.economist.com/news/science-and-technology/21731726-unimpressed-consumers-embrace-relevance-augmented-reality-instead-game (Accessed February 2018).

[2] "The History of Augmented Reality (Infographic)." Dennis Williams II. *Huffington Post.* March, 2016. https://www.huffingtonpost.com/dennis-williams-ii/the-history-of-augmented-_b_9955048.html (Accessed February 2018).

[3] Bitar Labs. http://www.bitarlabs.com

[4] Augment. http://www.augment.com

[5] "Escape reality with the best augmented reality apps for Android and iOS." Mark Jansen. *Digital Trends.* January, 2018. https://www.digitaltrends.com/mobile/best-augmented-reality-apps/ (Accessed February 2018).

[6] "Augmented Reality." Smart World. August, 2015. https://thayamora.weebly.com/smart-world/augmented-reality (Accessed February 2018).

[7] "Why do people keep giving Magic Leap money?" Adi Robertson. *The Verge.* October, 2017. https://www.theverge.com/2017/10/22/16505430/magic-leap-augmented-reality-temasek-funding-investment-why (Accessed February 2018).

[8] "Why do people keep giving Magic Leap money?" Adi Robertson. *The Verge.* October, 2017. https://www.theverge.com/2017/10/22/16505430/magic-leap-augmented-reality-temasek-funding-investment-why (Accessed February 2018).

CHAPTER 7

[1] "Marketing to Millennials? Make It Personal and Customized." Debra Kaye. *Entrepreneur.* July, 2014. https://www.entrepreneur.com/article/234891 (Accessed February 2018).

[2] "18 Crazy Facts About Bill Gates' $123 Million Washington Mansion." Madeline Stone. *Business Insider.* November, 2014. http://www.businessinsider.com/19-crazy-facts-about-bill-gates-house-2014-11 (Accessed February 2018).

[3] Vivek K. Singh, et al. "Motivating contributors in social media networks." *ACM Digital Library.* WSM '09 Proceedings of the first SIGMM workshop on Social media, 23 Oct. 2009, dl.acm.org/citation.cfm?id=1631149.

[4] "Plans Unveiled for Star Wars-Inspired Themed Resort at Walt Disney World." Jennifer Fickley-Baker. Disney Parks Blog. July, 2017. https://disneyparks.disney.go.com/blog/2017/07/plans-unveiled-for-star-wars-inspired-themed-resort-at-walt-disney-world/ (Accessed February 2018).

[5] "Massive Cantina and 'Star Wars' Hotel Coming to Disney World." Carlye Wisel. *Eater.* July, 2017. https://www.eater.com/2017/7/17/15984232/star-wars-cantina-bar-disneyland-disney-world-hotel (Accessed February 2018).

CHAPTER 8

[1] "Google Glass review: useful – but overpriced and socially awkward." Samuel Gibbs. *The Guardian*. December, 2014. https://www.theguardian.com/technology/2014/dec/03/google-glass-review-curiously-useful-overpriced-socially-awkward (Accessed February 2018).

[2] "Enhance Your World with a New Standard in Wearable Technology." https://epson.com/For-Work/Wearables/Smart-Glasses/Moverio-BT-200-Smart-Glasses-%28Developer-Version-Only%29/p/V11H560020 (Accessed February 2018).

[3] https://www.reconinstruments.com/products/jet/ (Accessed February 2018).

[4] https://www.accenture.com/us-en/success-airbus-wearable-technology (Accessed February 2018).

[5] "Aira uses smart glasses to help blind people navigate the world." Jessica Conditt. *Engadget*. January, 2017. https://www.engadget.com/2017/01/04/aira-blind-smart-glasses-phone-see-real-time/ (Accessed February 2018).

[6] "5 Educated Predictions for the Future of Amusement Parks." Sonia Weiser. *Mental Floss*. May, 2015. http://mentalfloss.com/article/64377/5-educated-predictions-future-amusement-parks (Accessed February 2018).

[7] "The Problem With Universal Volcano Bay's TapuTapu System." Josh Young. *Theme Park University*. October, 2017. http://themeparkuniversity.com/universal/

problem-universal-volcano-bays-taputapu-system/#Tou-
cA2EHD3YDBKcy.99 (Accessed February 2018).

CHAPTER 9

[1] "Sit back, relax, and enjoy a ride through the history of self-driving cars." Luke Dormehl and Stephen Edelstein. *Digital Trends*. January, 2018. https://www.digitaltrends.com/ cars/history-of-self-driving-cars-milestones/ (Accessed February 2018).

[2] "History of Autonomous Vehicles." https://sites.google.com/ site/socialinformaticsfinalproject/background-information (Accessed February 2018).

[3] "Uber's self-driving cars are now picking up passengers in Arizona." Andrew Hawkins. *The Verge*. February, 2017. https://www.theverge.com/2017/2/21/14687346/ uber-self-driving-car-arizona-pilot-ducey-california/ (Accessed February 2018).

[4] "Ten ways that driverless cars will change the world." Matthew Sparkes. *The Telegraph*. May, 2014. http://www.telegraph. co.uk/technology/google/10860036/Ten-ways-that-driverless-cars-will-change-the-world.html (Accessed February 2018).

[5] "Walt Disney World plans to deploy driverless shuttles in Florida." Russ Mitchell. *Los Angeles Times*. April, 2017. http:// www.latimes.com/business/autos/la-fi-hy-disney-shuttles-20170428-story.html (Accessed February 2018).

[6] "Driverless Cars could be heading to Walt Disney World." Erik.

Behind The Thrills. May, 2017. http://www.behindthethrills.
com/2017/05/driverless-cars-could-be-heading-to-walt-dis-
ney-world/ (Accessed February 2018).

7 "A Brief History of AI." *AI Topics*. https://aitopics.org/misc/
brief-history (Accessed February 2018).

8 "SOPHIA—The Audrey Hepburn inspired AI powered social
Robot !!" Vikas Jha. *Medium*. October, 2017. https://medium.
com/productivity-revolution/meet-saudi-arabias-new-
est-citizen-a-robot-named-sophia-ad4c37dd72a6 (Accessed
February 2018).

9 "5 Industries Being Most Affected By Artificial Intelligence."
Connie Chan. *The Fow Community Blog*. https://www.
fowcommunity.com/blog/future-work/5-industries-be-
ing-most-affected-artificial-intelligence (Accessed February
2018).

10 "5 extremely cool new AI applications." *Phrasee*. https://phra-
see.co/5-extremely-cool-new-ai-applications/ (Accessed
February 2018).

11 "Almost 90% of Theme Park Visitors Want Artificial
Intelligence, Virtual Reality to Be a Part of Their Experience."
Hospitality Technology. June, 2017. https://hospitalitytech.
com/almost-90-theme-park-visitors-want-artificial-intel-
ligence-virtual-reality-be-part-their-experience (Accessed
February 2018).

12 "Disney Planning to Introduce AI Robots into Their Theme
Parks." *AI Business*. March, 2017. https://aibusiness.com/
disney-planning-to-introduce-ai-robots-into-their-theme-

parks/ (Accessed February 2018).

[13] "AI and robots could threaten your career within 5 years." Marguerite Ward. *CNBC*. October, 2017. https://www.cnbc. com/2017/10/05/report-ai-and-robots-could-change-your-career-within-5-years.html (Accessed February 2018).

CHAPTER 10

[1] "'Tomorrowland' Disney History: Traveling Through The Theme Parks Past For A Glimpse Into The Future." Monica Castillo. *IB Times*. May, 2015. http://www.ibtimes.com/ tomorrowland-disney-history-traveling-through-theme-parks-past-glimpse-future-1929611 (Accessed February 2018).

[2] Pat Williams and James Denney. *How to be like Walt: capturing the Disney magic every day of your life*. Health Communications, 2004.

[3] Pat Williams and James Denney. *How to be like Walt: capturing the Disney magic every day of your life*. Health Communications, 2004.

Printed in Great Britain
by Amazon